Making Simple
Robots

Kathy Ceceri

MAKER MEDIA
SEBASTOPOL, CA

What You Need to Know 40
Gather Your Materials 40
Directions 41
Project: Design a Wheel-Leg Hybrid 48
What It Does 48
Where It Came From 48
How It Works 49
How Does a 3D Printer Work? 50
Making the Project 53
Project Parameters 53
What You Need to Know 54
Gather Your Materials 54
Directions 54

3. Unevolved Robots 73

Project: Make a Swarm of Gliding Vibrobots 77
What Is a Vibrobot? 77
What It Does 77
Where It Came From 77
How It Works 78
Making the Project 78
Project Parameters 78
What You Need to Know 79
Project: Make a Souped-Up Solar BEAM Wobblebot 83
What Is a BEAM Robot? 83
What It Does 83
Where It Came From 83
How It Works 84
Making The Project 84
Project Parameters 85

4. Robot Friends and Helpers 105

Project: Make a Chatbot Program 107
What Is a Chatbot? 107
What It Does 107
Where It Came From 107
How It Works 108
Making the Project 110
Project Parameters 112
What You Need to Know 112
Gather Your Materials 112
Project: Picture Yourself in the Uncanny Valley 124
What Is the Uncanny Valley? 124
What It Does 124
Where It Came From 124

How It Works 126
Making the Project 127
Project Parameters 127
What You Need to Know 127
Gather Your Materials 128

5. Fun, Artsy Robots **135**
Project: Make a littleBits Plotter 136
What Is a Plotter? 136
What It Does 137
Where It Came From 137
How It Works 139
Making the Project 140
Project: Make FiberBot, an E-Textile Arduino Robot 153
What Are E-Textiles? 153
What It Does 154
Where It Came From 154
How It Works 155
Making the Project 156

Afterword: What I Learned Writing This Book.............. **191**

Index **195**

Preface

I have always loved robots. When I was young, I devoured Isaac Asimov's stories, rooted for R2-D2 and C-3PO, and admired the cool stylishness of the replicants in *Blade Runner*. But I didn't try to build my own robots until I became an educator and a mom.

My son (the computer whiz) got his first Lego Mindstorms Robotics set when he was 12, and immediately set to work assembling his own creations. He attended robotics camp and learned to solder, work with metal, and program microcontrollers.

I watched his progress as a robot builder with envy … but as a writer and artist more comfortable working with cardboard and duct tape than software and hardware, I never thought about joining in.

Then two things happened. My family's homeschool science blog led to an invitation to join the writing community at the GeekDad site on *Wired.com* (and later to help create the GeekMom blog). And I was asked by an educational publisher called Nomad Press to write a children's activity book titled *Robotics: Discover the Science and Technology of the Future* (see Figure P-1).

I already knew a little about "real" robots from watching my son and interviewing robotics experts. And I now knew a bunch of knowledgeable folks I could call on for suggestions and advice. So I took the assignment, setting out to describe where robots came from and how they worked from the inside out. Because the book was aimed at 9- to 12-year-olds — and especially because I came to the topic with virtually no knowledge of electronics and mechanics — I wrote the book assuming my audience knew nothing as well. Starting from square one, I took readers on a tour of the various systems that go into robot design.

The projects I developed for that book presented a challenge in their own right. With a target audience of schools and libraries, Nomad set tight limits on how elaborate (and expensive) the do-it-yourself activities could be. As much as possible, the projects in *Robotics: Discover the Science and Technology of the Future* had to involve only ordinary crafts supplies and recycled materials. No kits, no soldering—not even a computer!

Figure P-1. *My first book of robotics projects was published by Nomad Press in 2012.*

Turns out, by finding ways to overcome those limitations, I became an expert in low-tech/no-tech robotics. And the response has been encouraging. In my workshops at schools, museums, and libraries, I see kids who have never taken a gadget apart before or wired up a motor to a battery become absorbed in designing their own moving robots. When I display my projects at Maker Faires and Mini Maker Faires, numerous parents and teachers tell me my projects and explanations speak to them at their level as well.

It's obvious to me that there are a whole lot of adults who would like to be able to do more with technology, robotics, and the Maker Movement without having to go back to school. It's easy to understand why: we are constantly surrounded by new technology that to most of us might as well be magic. Yet the news is filled with stories on why the public—especially kids—need to understand more about STEM (Science, Technology, Engineering, and Math) if we're to function in the world of the future. We want to know this stuff for our own satisfaction, and so we can help the children in our lives understand it too.

Yes, there are some great kits and beginner robot-building books out there. But I know there are still many adults who have more curiosity than experience with working with electronics. For many of us, even "basic" isn't basic enough.

That's why I wrote *Making Simple Robots* for Maker Media. I want to take those low-tech/no-tech projects that have proved so popular with people like me and push them just a *little bit* further. In this book I'll hold your hand as you tackle projects you might have shied away from before because of fear of the time, money, tools, skill level, or safety issues involved.

And I'm here to tell you, you don't have to go it alone. There are more opportunities than ever before for us nontechy types to mingle with engineers and veteran tinkerers, in the flesh and online. Here in upstate New York, I've been lucky enough to become part of a new local makerspace, the Tech Valley Center of Gravity in Troy (Figure P-2), where a host of knowledgeable people volunteer their time helping people like me get up to speed with the amazing new tools and materials around us. And don't forget the many educators, businesspeople, and hobbyists who generously share their expertise and suggestions on sites like *makezine.com* and *instruct-ables.com*.

Figure P-2. *The first home of the Tech Valley Center of Gravity in Troy, NY.*

If you've been frustrated, as I have, by experts who've forgotten what it's like to be a newbie and who fling jargon around that makes it seem like you'll never fit in, take heart. In *Making Simple Robots,* I've made sure that all the projects, regardless of their complexity, start at a level that feels familiar to anyone who's got basic arts and crafts experience. My hope is that *Making Simple Robots* encourages you to reach out and explore some of the newer technology taking over the hobby robot world.

If you're wondering how far a beginner can go with no formal electronics training, take a look at the Beatty family. It didn't take long for dad Robert (a mechanical engineer and retired cloud computing pioneer) to help daughters Camille and Genevieve, then 11 and 9, go from building kits to machining parts for their own robot designs. Beatty Robotics has created working models of space rovers for museums in New York and Prague. Here Camille, now 13, describes the path from novice robot builder to expert.

Where did you learn the skills to get started?

When I first came up with the idea of building robots, we didn't know anything. So, we started Googling around and learning what we could, then we bought some parts, and started trying to put things together. We watched YouTube videos to learn how to solder and we read lots of articles and blogs. We learned how to program Arduino microcontrollers, how to wire the electronics, how to machine metal, and all the other skills we needed. My dad learned along with us, but he gets us to do as much of the hands-on work as possible. After building our first six or so robots, we started to feel like we were getting the hang of it, but we always challenge ourselves with new designs.

How do you deal with frustration when things don't work?

Building robots is usually very fun, but, yes, sometimes it gets frustrating. Sometimes things don't work the way they should or you can't figure how to get something to work at all. Sometimes you get 90% done with something and then you break it or ruin it and you have to start all over again. That can be very frustrating. We have a whole box full of burned up electronics, broken tool bits, and other failures that we jokingly call the "Box of Shame." But we know that it's all part of the learning and experimenting process. A lot of times when we blow something up it's kind of fun. "Well, that's another one for the Box of Shame!" And then we try again. Other times, you think something up and you build it but it doesn't come out as good as you had envisioned. So, yes, there are many challenges, but you just keep pushing and learning and building more, and then when it comes out right, it feels really awesome. By the time we're done with a robot, we really love it, and don't want to let it go.

How do you keep track of what you are learning? Any shortcuts you can share?

First of all, we've learned to keep a detailed bill of materials on every robot we make, along with build notes and sometimes pictures. This way we can easily re-create any robot we've built. Our robots involve hundreds of parts, so we have found this to be very important.

Shortcuts we can share? We would advise that you build a little kit robot first to get your feet wet. Maybe build a couple of kit robots. You'll learn a lot of good skills. Then you start venturing off into your own unique designs. When you select your first kit, make sure it uses an Arduino microcontroller.

What is the best part of building robots for you? Is this something you (the girls) might want to go into as a profession?

For my sister and me, the best part of building robots is working with our dad. It's great fun and very rewarding. We learn everything together. Sometimes when something gets super technical or we run into a technical problem, our dad will figure it out on his own and then show us what he learned, but most of the time, we like to work together. Our dad has us do as much of the soldering, assembly, and machining as possible (regular school gets in the way, but our dad still makes us go!).

Both my sister and I have a lot of interests. Robots is just one of them. My sister loves drawing, writing, astronomy, and dress design. I love training and competing in dressage, and I also really enjoy photography, writing, and making jewelry. STEM is OK, but you add "Art & Design" and you get STEAM. For us, STEAM is where it's at. For us, building robots is not just science, but a creative art. Between my mom and dad and my sisters, we're always building stuff, and drawing stuff, and writing stories, and creating stuff. It's great fun for all of us.

Conventions Used in This Book

The following typographical conventions are used in this book:

Italic

Indicates new terms, URLs, email addresses, filenames, and file extensions.

`Constant width`

Used for program listings, as well as within paragraphs to refer to program elements such as variable or function names, databases, data types, environment variables, statements, and keywords.

`Constant width bold`

Shows commands or other text that should be typed literally by the user.

`Constant width italic`

Shows text that should be replaced with user-supplied values or by values determined by context.

This icon signifies a tip, suggestion, or general note.

This icon indicates a warning or caution.

Using Code Examples

This book is here to help you get your job done. In general, you may use the code in this book in your programs and documentation. You do not need to contact us for permission unless you're reproducing a significant portion of the code. For example, writing a program that uses several chunks of code from this book does not require permission. Selling or distributing a CD-ROM of examples from Make: books does require permission. Answering a question by citing this book and quoting example code does not require permission. Incorporating a significant amount of example code from this book into your product's documentation does require permission.

We appreciate, but do not require, attribution. An attribution usually includes the title, author, publisher, and ISBN. For example: "*Making Simple Robots*, by Kathy Ceceri (Maker Media) Copyright 2015, 978-1-4571-8363-8."

If you feel your use of code examples falls outside fair use or the permission given here, feel free to contact us at *bookpermissions@makermedia.com*.

Safari® Books Online

 *Safari Books Online* is an on-demand digital library that delivers expert content in both book and video form from the world's leading authors in technology and business.

Technology professionals, software developers, web designers, and business and creative professionals use Safari Books Online as their primary resource for research, problem solving, learning, and certification training.

Safari Books Online offers a range of plans and pricing for enterprise, government, education, and individuals.

Members have access to thousands of books, training videos, and prepublication manuscripts in one fully searchable database from publishers like O'Reilly Media, Prentice Hall Professional, Addison-Wesley Professional, Microsoft Press, Sams, Que, Peachpit Press, Focal Press, Cisco Press, John Wiley & Sons, Syngress, Morgan Kaufmann, IBM Redbooks, Packt, Adobe Press, FT Press, Apress, Manning, New Riders, McGraw-Hill, Jones & Bartlett, Course Technology, and hundreds more. For more information about Safari Books Online, please visit us online.

How to Contact Us

Please address comments and questions concerning this book to the publisher:

> Make:
> 1160 Battery Street East, Suite 125
> San Francisco, CA 94111
> 800-998-9938 (in the United States or Canada)
> 707-829-0515 (international or local)
> 707-829-0104 (fax)

Make: unites, inspires, informs, and entertains a growing community of resourceful people who undertake amazing projects in their backyards, basements, and garages. Make: celebrates your right to tweak, hack, and bend any technology to your will. The Make: audience continues to be a growing culture and community that believes in bettering ourselves, our environment, our educational system—our entire world. This is much more than an audience, it's a worldwide movement that Make: is leading—we call it the Maker Movement.

We have a web page for this book, where we list errata, examples, and any additional information. You can access this page at *http://bit.ly/making_simple_robots*.

Acknowledgments

Thanks to the team at Maker Media—Brian Jepson, Patrick DiJusto, Frank Teng, and of course Dale Dougherty—for their help and encouragement in creating this book. I'd also like to acknowledge the support and inspiration of the following people and companies in my pursuit of robot-building ideas, skills, materials, and information:

- Robert, Camille, and Genevieve Beatty (*http://beatty-robotics.com*)
- John Rieffel at Union College (*http://muse.union.edu/crochetlab/*)
- Michael Dickey, Jan Genzer, and Ying Liu at North Carolina State University (*http://www.che.ncsu.edu/dickeygroup*)
- The Lifelong Kindergarten Group at the MIT Media Lab for the use of Scratch (*http://scratch.mit.edu*)
- NASA Ames Research Center (*http://magicalrobot.org*) and Vytas SunSpiral
- George M. Whitesides and his research group at Harvard University (*http://gmwgroup.harvard.edu*), especially Filip Ilievski
- Hiroshi Ishiguro of Osaka University (*http://eng.irl.sys.es.osaka-u.ac.jp*)
- Saul Griffith at Otherlab (*http://otherlab.com/*)
- Jie Qi (*http://technolojie.com*) of MIT and Circuit Stickers (*http://chibitronics.com*)
- Ben Finio (*http://sciencebuddies.com*)
- littleBits (*http://littleBits.cc*)
- Nick Kohut at Dash Robotics (*http://dashrobotics.com*)
- Solarbotics (*http://solarbotics.com*)
- Becky Stern at Adafruit Industries (*http://adafruit.com*)
- Windell H. Oskay of Evil Mad Scientist Laboratories (*http://evilmadscientist.com*)
- SparkFun (*http://sparkfun.com*), especially Jeff Branson of the Department of Education
- Strandbeest (*http://strandbeest.com*) artist Theo Jansen
- Marc and Emma Edgar (*http://bit.ly/1mxh4W1*)

And all the folks at the Tech Valley Center of Gravity makerspace in Troy, NY (*http://techvalley centerofgravity.com*)!

Introduction

Robots are getting simpler all the time.

If that statement sounds backwards, consider this: as we continue to cram ever-more-powerful electronics into tinier and tinier containers, the number of things we can do at home, ourselves, with our everyday appliances and devices is growing exponentially. The same technology that makes it possible to squeeze a GPS, tilt sensor, camera, and wireless Internet connection into your cell phone also lets researchers design miniaturized robots that really pack a punch.

For scientists, knowing that a robot brain doesn't need to take up more room than a postage stamp means they can design robot bodies out of the most unexpected materials: soft polymers, folded paper, fabric that can bend and stretch instead of breaking. At one time building a robot out of such materials might have sounded nuts. But today even high schools and hobbyists, not to mention universities and commercial labs, have access to tools like 3D printers and laser cutters that can produce lightweight robot bodies able to survive falls from the tops of buildings —or from outer space. These new robots are more resilient and yet easily replaceable. If one prototype doesn't work, it's easy to tweak the design and cobble together another.

It's not just robot bodies that are getting simpler, though. Roboticists are taking a look around them at the natural world, especially lower-order animals that exhibit complex behavior without a lot of what we would traditionally think of as "smarts." Where once the goal was to create a humanoid device that could do what people do, now the trend is to make something small yet effective. Instead of a giant laser-eyed Gort clomping across the landscape or Data with his positronic brain, we're getting the skittering mechanical spiders of *Minority Report*.

Roboticists are also asking whether a robot needs to be able to avoid every misstep and obstacle if it can easily pick itself up and keep going. A robot that can travel along through pretty much anything using only simple rhythmic motions can make do with a lot less computing power than one that needs to decide where to place every footstep based on sensory readings of the surrounding terrain. This frees up processing capability that can be devoted to higher-level tasks. It can also lead to pared-down, interdependent swarms of micro bots that do more as a team than they could do alone. The hive mind is real. And modular robots that can assemble and repair themselves are already on the horizon.

For people who like to learn about robots, but who don't have a background in science or technology, this is all great news. We watch in fascination as researchers turn old bicycle tires and PVC pipes into robotic arms and autonomous rovers. And as robots get simpler, we can begin to build our own designs that share many of the traits of "real" robots without a fancy lab or an engineering degree. When constructing a robot body takes nothing more advanced than zipties and a sewing machine, you're now in a place where even rank beginners can get into the act.

That's where this book comes in. *Making Simple Robots* will show you how to use standard crafting techniques and skills to build designs that are only a few degrees away from the real thing. Don't worry about getting in over your head. Many of the projects in *Making Simple Robots* will introduce you to some aspect of robotics without requiring you to build an entire machine. Some can be done in one session. Others can easily be broken down into steps that can be completed over a few afternoons.

If you've ever felt intimidated by a beginning robotics book, *Making Simple Robots* is for you. None of the projects in this book assume any background in electronics. Unfamiliar tools and techniques will be described in detail.

However, this book is also aimed at getting you out of your comfort zone. Curious about 3D printing? You'll find a project to help you ease into it. Wondering how an Arduino microcontroller (a miniature "computer on a board" designed to run lights, motors and other electronics) works? You'll get a chance to find out.

What Is a Simple Robot?

The traditional definition of a robot is a machine that can sense, think, and act. That means it can:

- Tell what's happening in its physical environment
- Analyze that information and make a decision based on its programming
- Move around, flash lights, sound alarms, or otherwise do something that affects the physical world

Think of it this way: your standard upright vacuum turns on when you flip a switch, but it will sit in one spot forever unless you grab the handle and start pushing it over the carpet. Banging it into a chair leg (or a dog's tail) doesn't affect its operation at all—although you may change your actions accordingly. A Roomba, on the other hand, is a vacuum with a brain. It can be programmed to turn itself on at certain times. And when it bumps into something, it backs up on its own and goes around it. You may think of a Roomba as a glorified appliance, but it's actually a full-fledged robot. In fact, the iRobot company that makes the Roomba also supplies combat-ready, autonomous search-and-rescue robots to the military.

The "sense, think, act" definition is useful when you're trying to explain the difference between an appliance and a robot. But what is considered a robot has been changing in recent years, along with advancements in technology. For one thing, embedded sensors and processors are turning practically every device we use into a type of robot. The automatic garage door with the safety sensor that stops the door from closing if something or someone is in the way—that's a robot. So is the dryer that stops tumbling when its moisture-sending electrodes tells it the towels are done.

But more importantly, a new way of approaching the field of robotics is considering how to build "dumb" machines and structures that still behave in sophisticated ways. Primitive robots that use switches as sensors—such as the push button that reverses the motor on a bump-and-turn robot—have been around for a while. The area of BEAM robotics (an acronym for Biology, Electronics, Aesthetics, and Mechanics) makes use of the fluctuating power from solar panels to create lifelike unpredictability. But new research is looking at whether the design of a robot body that determines a machine's behavior itself, separate from any electrical controls, can be considered a form of programming.

This book uses the word "robot" in its broadest sense. That includes robotic bodies and parts that are not necessarily electronic or even motorized. Some projects are conceptual, and don't necessarily involve any building at all. But all are based on actual robotics research being done today.

What's in This Book?

Making Simple Robots is divided into chapters, each focusing on an interesting topic in robotics. You'll get a look at new developments, and the surprising ways simple robots are being used in applications from sheer fun and entertainment to some of the most advanced research being done today. And you get to make related projects of your own. Here's how they're broken down:

Chapter 1, *Robots Made from Interesting Materials*

This is a look at some of the state-of-the-art materials and manufacturing processes that are opening up new possibilities in robot construction.

Chapter 2, *Robots That Get Around*

One of the most interesting challenges in robot design is figuring out how to get from place to place. Learn about some of the unusual ways researchers are solving the problem of robot locomotion.

Chapter 3, *Unevolved Robots*

Take a look at vibrating robots, BEAM robots, and other primitive creations that stretch the definition of "intelligence."

Chapter 4, *Robot Friends and Helpers*

Social robots just want to be loved. This chapter looks at the challenges involved in making helper robots look and act friendly, not threatening.

Chapter 5, *Fun, Artsy Robots*

Robots are useful, but they're playful, too. Discover some of the ways that artists, toymakers, and educators are using robotics to add life and excitement to their designs.

How Are Projects Laid Out?

The projects in this book are intended to enhance your understanding of robots and robot-building techniques. The descriptions introduce you to the type of working robot or robotics research that inspired the project. The directions cover the things you need and the things you'll do in great detail. By following them closely, you should be able to successfully complete the project as described. But the best way to learn something is to make it your own. So many of the projects go further, offering suggestions for adapting the project for different abilities and settings. They may also include ideas for extending the project to dive more deeply into the topic, or to increase the complexity for those looking for more of a challenge. There's even advice for making some projects the basis of an exhibit at a science fair—or a Maker Faire.

Developing any Maker project, from the simplest craft to the most complex electronics, involves planning, preparation, and documentation. So to give you a peek into the process of creating a project from scratch, woven in and around the directions for building each project you'll find explanations of the behind-the-scenes steps involved in creating it. This format was inspired by a joint talk given at World Maker Faire New York 2013 by Tech Valley Center of Gravity member Steve Nordhauser and the author. This added information will help you understand the project in greater depth, and be useful if you decide to create your own variations.

The project descriptions format is as follows:

- What Is a _____?
- What It Does
- Where It Came From
- How It Works
- Making the Project
  — Project Parameters
    — Time Needed
    — Cost
    — Difficulty
    — Safety Issues

- What You Need to Know
    - Skills You Already Have
    - Skills You Will Learn
- Gather Your Materials
- Directions

 Don't forget to document your work!

This reminder can save you from repeating the same mistakes the next time you try a similar project. Keep a record of your materials and what you do so you can refer to them whenever you need to. In high school and college science classes and in research laboratories at universities and businesses, a permanently bound lab notebook in your own handwriting, with no erasures or torn-out pages, can provide proof that your invention or discovery is really your own. For that purpose, a hardbound notebook like the Maker's Notebook from Maker Shed (*http://www.makershed.com*) is ideal. Some tips:

- Start each work session on a fresh page. Cross out any blank spaces.
- Write neatly so you can decipher your notes later.
- Write down every detail as it happens. Don't trust to memory.
- Write down names and contact information for anyone you consult with. Make note of any books or websites you consult.

While cultivating the habit of keeping a standard laboratory notebook is a worthwhile goal, beginners and die-hard hobbyists have other options for documenting their work. One quick and easy way is to take photos or videos as you go along. Be sure to jot down enough notes to help you remember what the pictures are showing. If your project is new and different and your documentation is clear and easy to understand, you can share your process with others on a site like Instructables. For additional ways to show off your creations with other *Making Simple Robots* readers, visit Crafts for Learning.

The directions for each project are organized into the following steps:

Step 1: List Your Requirements

What do you want your robot to do? Here's where you describe its function; suggestions for how you'll accomplish that come later.

Step 2: Plan Your Project

Consider the options for building the project, and choose which ones you'll work with.

Step 3: Stop, Review, and Get Feedback

Before you begin building, go back over your plans one more time. If you've got questions, now's the time to look them up in this book, in other resources, or to ask for advice from more experienced builders. Helpful hints will be included in this section.

Step 4: Build Your Prototype

Actual building instructions are listed here.

Step 5: Test Your Design

Once you've assembled your robot project, you'll need to try it out. Here's how to see if it's working according to expectation.

Step 6: Troubleshoot and Refine

Look here for suggestions to fix some common problems that may arise in your design.

Step 7: Adaptations and Extensions

The directions in this book are just a jumping off point for creating robots of your own. This section may include ideas for adapting the project for young kids, for a classroom or group setting, or for a research project.

Ready? It's time to start Making Simple Robots!

Other Handy Information

Every now and then, you may see a little sidebar or box labeled "Break the Code" with trivia or inside jokes related to the aspect of robotics being discussed. These factoids can help newbies begin to sound knowledgeable quickly and easily.

Throughout the chapters you will also find Linkboxes containing the addresses of useful websites as they are mentioned.

For updates and news about *Making Simple Robots*, visit Crafts for Learning (*http://www.craftsforlearning.com/makingsimplerobots*).

Robots Made from Interesting Materials | 1

When you think of robots, what do you picture them being made out of? In classic science fiction books and movies, it's almost always sheet metal. Household robots and robotic toys are usually made of hard plastic. But in robotics labs around the world, scientists are looking at every material imaginable. Today state-of-the-art materials and manufacturing processes are opening up new possibilities in robot construction. Instead of using heavy, rigid bodies, researchers are trying out thin, flexible skins that let robots bend, squeeze, and stretch. One goal is the creation of robots that mimic their biological counterparts. These biomimetic robots move more like living things, and often need less programming and less power to get them going. They are also usually more "compliant"—a robotics term that means they yield when they run into people and things in their path, instead of mowing them down.

Even more exciting is the idea of a "smart body" that uses the overall design or molecular structure of a robot to control how it behaves. One way is by using programmable material such as *shape memory* polymers and alloys. These are materials that can be trained to change their physical size or shape when exposed to an outside stimulus like light or heat. For instance, a straight piece of shape memory alloy (SMA) wire can be preset to coil up like a spring when placed in a cup of hot water.

Some researchers are working on smart bodies made of a common childhood art material—shrinkable plastic (similar to those sold in crafts stores under the name Shrinky Dinks). When heated, the plastic sheets contract to about 60 percent of their original height and width. Kids draw on them and crafters run them through computer printers to make colorful pendants. But at North Carolina State University (Figure 1-1), researchers printed thick black lines on the sheets and exposed them briefly to infrared light, the kind used in heat lamps. Since dark colors absorb more light energy than light colors, the black lines heated up while the clear parts stayed cool, causing the sheets to bend. The results were self-folding sheets of plastic that could transform themselves into boxes or other 3D shapes automatically.

Figure 1-1. *Self-folding sheets of shrinkable plastic that are activated using heat. Credit: Ying Liu and Jacob Thelen, North Carolina State University.*

However, Shrinky Dinks can't go back and forth between their shrunken and stretched-out states. Other shape memory materials can. Researchers are using this flexibility to create artificial muscles, which fall into the category of *actuators*, things that make a robot move. In 2009, Ray Baughman at the University of Texas at Dallas showed off artificial muscles made of a type of shape memory polymer called carbon nanotube. The atoms in a carbon nanotube are arranged in a honeycomb, like the surface of a geodesic dome. This makes the material as stiff as steel in one direction, but stretchy in the other. It can also operate at extreme temperatures, which makes it ideal for space missions. Baughman's team originally used the material in the form of an aerogel, sometimes described as "frozen smoke." Aerogel is a solid so light that it has virtually the same density as air. A voltage applied to a strip of carbon nanotube aerogel made it expand 10 times farther and 1,000 times faster than a natural muscle. More recently, Baughman created wax-filled strands of carbon nanotube yarn that expanded when the wax was melted at high temperatures. Someday the yarn may be woven into smart materials that change shape when armed or cooled.

SMA wire made out of nitinol—a blend of the metals nickel and titanium—is being used as an actuator on very small and very flexible robot bodies. In 2012, the Octopus Project released video of an underwater robot using its SMA-actuated tentacles to crawl across a wading pool. And at Virginia Tech, researchers are working on Robojelly, a jellyfish-shaped robot that moves with the help of strips of SMA composite. The Robojelly helps engineers and biologists study how jellyfish swim without having to worry about keeping a real animal alive and healthy. The hope is to create a design that can someday be applied to underwater vehicles.

Along with new materials, many forward-thinking inventors are also turning to the past for inspiration. Among the avenues they are exploring are low-tech materials like paper and rubber. Their light weight and springiness can make them easier to power. And because they're so commonly available, they're easier and cheaper to design, build, upgrade, and replace. In fact, some researchers believe they can be manufactured in such large quantities as to become disposable—send them out to explore or do a job and don't worry about whether they can make it back.

The projects in this chapter will let you get your feet wet making simple robots by playing around with some of the interesting materials being used in cutting-edge robotics research today: shape memory alloy wire, paper, and rubber. The projects themselves are extremely basic —more like proof-of-concept designs that test the capabilities of the materials than something resembling an actual robot. But the Adaptations and Extensions section of the project descriptions point you towards further areas of exploration if you want to keep going. Be sure to check the Linkboxes for additional background information and tutorials online.

Project: Make Actuated Paper

Actuated paper is a type of programmable material that can move on its own with the help of embedded shape memory material.

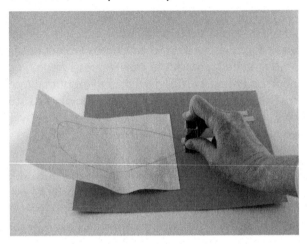

What It Does

Actuated paper can fold and refold itself into various shapes as needed. Researchers are studying ways to use it for all-purpose robots that can alter their shape as needed. Transformable robots would be very handy on space missions and other places where getting new supplies could be difficult. Artists and designers are also using actuated paper to make kinetic sculptures and furnishings. By adding sensors and other electronics, it's possible to create actuated paper structures that respond to the environment around them.

Where It Came From

Paper was invented by the Chinese around 105 AD, and it opened up a whole new world of possibilities in what we now call product design. As a surface for print, it replaced parchment and papyrus for books and metal for money. And because it was thin enough to fold, stiff enough to hold its shape, and light enough to be portable, it was also useful for boxes, containers, toys, and decorations. By the 1700s, mechanized papermaking brought the price down enough for ordinary people to afford, and folding and cutting paper into designs became a popular hobby in Europe and Asia.

The use of paper as a material for robot bodies takes its inspiration from two popular artforms: pop-up illustrations and origami. Pop-up illustrations, used in children's books and greeting cards, are folded structures that lie flat when the pages are closed, and spring upright when the pages are open. Origami, the traditional Japanese art of paper folding, is both a favorite children's pastime and a high art. Extremely complex and beautiful designs can be created by using techniques like modular designs (building a larger piece out of smaller separate pieces of origami) and technical folding (pre-creasing the paper in regular geometric patterns). Action origami pieces, which move when you pull or push certain spots, range from the traditional hopping frog to amazing spinning spirals and tumbling boxes. Origami designers share their original creations using written directions, diagrams, or crease patterns that indicate where the paper was folded on a flattened piece of paper.

In recent years origami engineering has become a field of scientific study. In 2009, MIT mathematician Erik Demaine and a team at MIT proved that a type of origami fold called the box pleat pattern could be used to create basically any shape possible. Next, Demaine, MIT roboticist Daniela Rus, and Harvard's Robert Wood created a box pleat–creased sheet actuated with SMA and magnets that could fold itself into a boat and a paper airplane. Then in 2014, the same team published an article describing a heat-activated origami robot made from a flat sheet made of layers of copper, paper, and shape memory polymer.

Figure 1-2. *This robot made of paper and heat-activated shape memory polymer from Harvard and MIT scientists can fold itself up and scurry away. Credit: Seth Kroll/Harvard's Wyss Institute.*

The copper inner layer was etched with traces that could carry electrical current. When a battery was connected, an attached microprocessor would send current through the copper, heating up the polymer, which would begin to fold up in sequence. Such a robot could be programmed to take on different shapes. One demonstration model that folded itself into the shape of a bug could immediately scurry away on four motorized legs.

Other researchers like Jie Qi of the MIT Media Lab are bridging the gap between origami engineering and art by making cranes that flap their wings and paper "vines" that retract when touched (see Figure 1-3).

Figure 1-3. *Jie Qi's moving paper sculptures are activated by shape memory alloy wire. Credit: Jie Qi.*

Paper folding and pop-ups are also being used to create biomimetic robots. Researchers at the UC Berkeley Biomimetic Millisystems Lab build tiny robots from folded cardboard using a process they call Smart Composite Microstructures. A flexible layer of plastic is sandwiched between two layers of cardboard. When the robot model is folded into shape along thin precut slits, the plastic inner layer become a flexible hinge that is nearly frictionless. Assembling a working robot using this process takes less than an hour, and the use of inexpensive materials makes it possible for researchers to continually refine the design and create new prototypes to test out.

One of the first minirobots built using this approach was an open source matchbox-sized insectoid called RoACH (Robotic Autonomous Crawling Hexapod). RoACH is actuated by shape memory alloy wires that pull a sliding plate attached to its six legs. RoACH has two *degrees of freedom*, or directions in which a part of its body can move. Its sliding plate can move up and down to lift and lower the legs, and back and forth to make the legs sweep backward or forward. That's all it needs to make its legs propel it forward or backward. A later design, DASH (Dynamic Autonomous Sprawled Hexapod), uses a motor to drive the six legs. DASH can run along at 15 times its body length per second and bounce back from falls the height of a multistory building. It has even been adapted into a Dash Robotics DIY kit for students and hobbyists that can be controlled by a smartphone or tablet (Figure 1-4).

Figure 1-4. *Dash is a robotic bug made from laser-cut cardboard and easily assembled. Credit: Dash Robotics.*

How It Works

As a material for building robots, paper is both stiff enough to stand up and support a small load, yet soft enough to act like a spring when folded. Paper robots like DASH and RoACH use a kind of pop-up hinge called a Sarrus linkage, which opens into a hollow parallelogram shape that can swing back and forth to create a walking motion. Action origami models like the hopping frog rely on the potential energy created when paper is folded and held down under pressure.

Nitinol was discovered in 1959 by Navy researcher William J. Buehler. (Its name stands for Nickel Titanium Naval Ordnance Laboratory.) Extremely flexible and strong, it has been used in orthopedic and cardiovascular surgery, solid-state heat engines, pipe couplers for aircraft, eyeglass frames, and toys. Its shape memory comes from the fact that the atoms in each nitinol molecule rearrange themselves into different crystalline shapes at different temperatures. At high temperatures, the atoms go into what is called an *austenite phase*, in which each nickel atom is surrounded by eight titanium atoms arranged in a cube. At lower temperatures, they revert to a *martensite phase*, a more complex arrangement of atoms. There are various kinds of nitinol, which can be programmed to phase shift at different temperatures for different uses. Some nitinol products can be programmed by the user. Other kinds come preset.

Robot builders often use Flexinol, a brand of nitinol wire made by Dynalloy. It's preset to shrink between 5 and 10 percent when heated by an electrical current. When cool, it can be stretched back out to its original resting size. Because the change in size is so small, and because it needs some kind of force to help it stretch back out to its original size, the effect of a Flexinol actuator depends on the design of the robot and what it is made of. If it is anchored to a stiff material like metal or wood, a spring or rubber band can be used to pull it back to its resting length. In a paper model, gravity or the springiness of the paper itself can serve the same purpose. However, in simple paper models, the movement created by a Flexinol actuator is usually very subtle. It's great for making paper sculptures wave gently in a lifelike motion. Really dramatic effects, such as robobugs that run or fly, require more advanced paper engineering.

Making the Project

When it comes to building robots, you can't get simpler than folded paper. For this project, you won't be building a robot that walks and talks—but you will get to practice the kind of rapid prototyping techniques used in real robotics labs by designing a smart body using common household materials. You'll also get to practice making an electrical circuit, and discover how to use the futuristic programmable material Flexinol to make actuated paper that jumps, sways, and waves.

While you can find almost all the supplies for this project around your house or at your local crafts or hardware store, you will most likely have to order your nitinol wire online. Be sure to get extra, since without a controller or protective components there's a slight chance of frying your Flexinol. More importantly, you'll want enough to play around with as you come up with your own ideas for building actuated models from paper and other materials.

 Don't forget to document your work!

Project Parameters

- Time Needed: 2–3 hours
- Cost: $15–$40
- Difficulty: Easy to moderate

- Safety Issues: Some 9V batteries come "overcharged" from the factory and may deliver too much voltage at first. Be careful to use batteries in very short bursts so as not to overheat the SMA wire or create sparks around the paper model.

What You Need to Know

- Skills You Already Have: Craft skills (sewing, beading) helpful but not required
- Skills You Will Learn: Building an electrical circuit

Gather Your Materials

- Flexinol wire, at least 2 feet long (note: I used .008-inch high temperature (HT) wire for this project. You may want to start with .006-inch diameter, which according to the manufacturer can be activated continuously without burning out. A 5-meter (15-foot) roll is less than $25 from RobotShop (*http://www.robotshop.com*)).
- Masking tape
- Wire cutters
- Needle-nose pliers
- Two or more small metallic jewelry crimp beads (available in packs of 50 at craft stores)
- Peel-and-stick aluminum foil tape (look for a duct-tape-sized roll of actual foil, available in the heating department of hardware stores, not metallic colored duct tape)
- Pen
- Scissors
- Paper—letter-sized copy paper and/or cardstock
- Clear, wide packing tape
- 9V battery (nonrechargeable)

Directions

Step 1: List Your Requirements

The goal of this project is to explore different ways to use the springiness of folded paper in robot bodies and other moving models, and how to magnify the movement provided by the Flexinol wire actuator.

Step 2: Plan Your Project

There are several different paper building techniques you can try with the Flexinol wire. The simplest and most dramatic movements are making a strip of paper curl up, or creating a hinge to open and shut a flap of paper. To avoid wasting SMA wire, you can build a test platform that lets you try out multiple paper designs using the same piece of wire.

Step 3: Stop, Review, and Get Feedback

Before you begin, you might want to research more about origami techniques. Practice interesting models and learn about the traditional and technique bases used to create new designs by checking out the websites in the Linkbox below, including YouTube how-to videos by instructors like Jeremy Shafer, or any of the books by Robert J. Lang.

For more background about nitinol and how it works, the Dynalloy and Robotshop websites, also linked below, contain useful information. The Make: website also has a detailed article by MIT Media Labs' Jie Qi, and you'll find inspiring actuated paper art on her website.

Only activate the wire for a few seconds at a time—just until the wire stops contracting. Overheating the wire can "deprogram" it—or worse, set the paper on fire.

Step 4: Build Your Prototype

Follow these steps to build a test platform for your SMA wire:

1. Use the wire cutters to clip a piece of .008-inch diameter Flexinol wire that is 18 inches long. (If you are using .006-inch wire, you can use a piece as short as 9 inches.) For ease of handling—and to avoid losing track of the hair-thin, bouncy wire until you attach it to something—wrap a bit of masking tape around it like a tag.

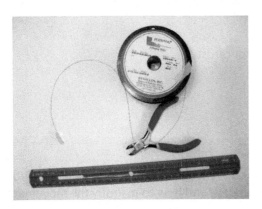

2. The wire will be easier to work with and make a better electrical connection if you crimp the ends. (Crimping is also necessary if you want to solder SMA wire, since solder will work its way off of the wire itself as it expands and contracts.) Take the pliers and bend about 1/4 inch (6 mm) of one end of the wire into a tight U shape. Slip a crimp bead onto the wire and around the bend. Then push the bead back, this time catching the loose end in the bead as well.

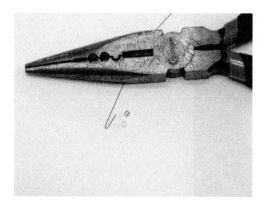

3. Crunch down on the bead with the pliers, squeezing as hard as you can until the bead is flat. Check whether the bead can still slide up and down—if it can, angle the pliers a bit and squeeze again. Crimp the other end the same way.

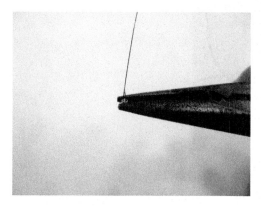

4. Next, you will use narrow strips of peel-and-stick adhesive aluminum foil tape as hand-made circuit wiring, known as *traces*. (Thanks to *Make:* magazine writer Chris Connors for this trick!) Cut a piece of foil tape about 6 inches long.

Flip it over to the paper side, and use a pen to mark off five long strips each 3/8 inch (1 cm) wide. Cut along the lines.

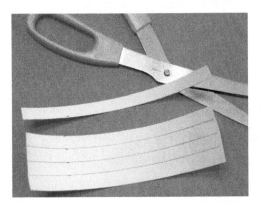

5. Take one of the narrow strips of foil you just made and cut it into two pieces, one 3 inches (8 cm) long and the other 2 1/2 inches (6 cm) long. Take one of these short pieces and peel off about half an inch (1.25 cm) of the backing. Now take one of the crimped wire ends and press it onto the exposed foil.

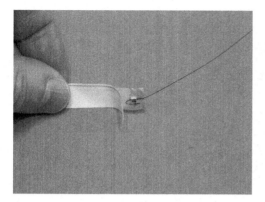

Close up the backing and squeeze to make a good connection between the wire and the foil. Do the same with the other short piece of foil and the other crimped wire end.

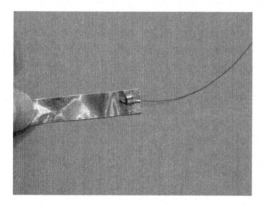

6. To make the base of your test platform, take a piece of letter-sized paper or cardstock and place it on the table in front of you, portrait-style. With the pen, make two marks side by side, 1/8 inch (3 mm) apart, just below the center of the page. Take one of the the foil strips holding the wire, and unpeel the flap of backing you exposed before. Press it onto the paper and slowly unpeel the rest of the back so that it does the same with the other piece of foil, placing it parallel to the first strip and leaving a small gap in between them. It's OK if the loop of wire hangs off the top of the paper.

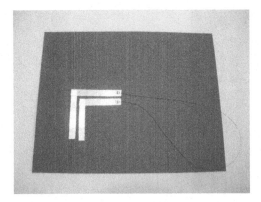

7. If you like, extend the circuit by continuing your double track of tape around the corner. Cut another strip of foil into two pieces, both the same size. Take one new piece and place it on top of the end of one of the first foil strips at a right angle. Peel off the backing and press hard to help secure the connection between the top and bottom pieces of foil. Do the same with the other piece of foil, again being careful to keep a gap between the tracks. You can continue on this way as far as you like—but be aware that glued connections like these are not always reliable. The tail end of a long winding foil trace may end up being more decorative than functional. The section closest to the SMA wire will still be usable, regardless.

8. To protect the test platform from tearing when you attach and remove paper models, cover the upper area with a protective layer of tape. Start right above the foil strips to the top of the page. Wide, clear packing tape is best, although masking tape works, too. Just lay down strips of tape, one next to the other.

Step 5: Test Your Design

Before you start trying out different paper models, preheat the wire to make sure it's working and let it stretch out a bit. Tack down the loop of wire to the test platform with several narrow strips of masking tape "bandages." You can make these easily by unrolling a length of tape, cutting a fringe along one edge, and pulling off strips as needed. To use, press a bandage down across the wire, putting a little less pressure on the middle than on the ends, so the wire can move underneath. The masking tape bandages can be reused a couple of times to save time while you're working. (You can also use a loop or tube of masking tape, sticky side out, to attach a paper model to the cardstock test platform.)

Make sure to secure the wire where the ends come together at the bottom, to prevent them from touching each other when they start to move.

Then take the 9V battery and briefly touch the terminals to the foil tracks. You should see the wire try to twist and curl. Lift the battery back up as soon as the movement is done—no longer than three or four seconds at a time. Too long and you risk overheating the SMA wire, which will de-program its shape-changing ability. When you're ready, try actuating some paper models. Here are two simple examples:

Curling Flap

1. Take a piece of letter-sized copy paper and cut it in half, so that you end up with two pieces 8 1/2 by 5 1/2 inches (22 by 14 cm). (Save the other half for the next model.) Make a sticky-side out tube of masking tape and use it to attach the strip of paper near the bottom of the testing area of the platform, close to the foil. The flap should lay flat. It's OK if it hangs off the top.

2. Lay the loop of SMA wire on top of the paper flap. Attach the wire to the flap with the little masking tape "bandages." Place them across the wire, about 1/2 inch (1.25 cm) apart, closer around the top of the loop.

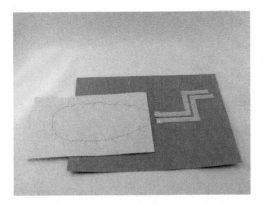

3. Touch the 9V battery to the foil traces. The flap should curl up, but it may also bend in a wave. You can change how it moves by playing around with the placement and number of the masking tape strips. You can also try creasing the flap in one or more places to see how that changes the movement of the paper.

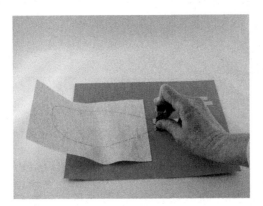

Paper Robotics Linkbox

Makezine Shape Memory Alloy How-To (*http://bit.ly/mem-alloy*)

Jie Qi (*http://technolojie.com/*)

Robert J. Lang (*http://www.langorigami.com*)

UC Berkeley Biomimetic Millisystems Lab (*http://bit.ly/ucb-biom*)

DASH Robotics (*http://dashrobotics.com/*)

Jeremy Shafer's YouTube channel (*http://bit.ly/jeremy-shafer*)

Add a closing paper mouth to your project

 In origami-speak, "valley folds" go down and "mountain folds" go up.

1. Start with a piece of copy paper 8 1/2 by 5 1/2 inches (22 by 14 cm). Place it portrait-style in front of you, and fold it in half by taking the bottom edge and bringing it to the top. Then take the upper edge and fold it down towards you about 3/4 inch (2 cm) from the top. Unfold it. Fold down one corner so that it lines up with the crease. Do the same with the other corner. Then flip the whole thing over and do the same with the other edge. Adjust the edges so that they stick out at right angles from the rest of the paper. These are the "lips" on the mouth.

2. Turn the paper so that the fold is on the top. Fold it over again towards you about 1/2 inch (1.25 cm) from the top. Without unfolding it, flip the whole thing over and fold it over towards you one more time, the same amount. Unfold all the folds you just made

and open the paper up so that the sides are pointing up in a crooked V shape. Adjust the creases so they go valley-mountain-valley-mountain-valley, like an accordion.

3. Close the model up and lay it on one side. You now have a paper "mouth" that wants to snap shut when you open it. Add two "eyes" by taking the top corners and folding the tips of them back up, so little triangles of paper are peeking out behind the diagonal folded corners.

4. Form a tube out of a piece of masking tape, sticky-side-out, and put it along the bottom edge of the paper model, near the accordion fold. Attach the taped edge to the test platform so the mouth is facing away from the foil.

5. Use the thin masking tape bandages to attach the SMA wire loop to the top of the paper model. The wire should go up along the side edges, and across the top of the fold (on the outside). Tack down any extra wire to the testing platform.

6. Turn the testing platform around so the mouth is facing you. Take the 9V battery and touch it to the foil tracks. The mouth should slowly open. Lift the battery after

a few seconds, and the mouth will slowly close, thanks to the springiness of the accordion fold.

Step 6: Troubleshoot and Refine

For maximum effect, keep the wire as taut as possible. If it stretches and loosens up, gently pull it down towards the foil traces, being careful not to break it. Then move the paper model itself further back towards the top of the platform, until the wire is as tight as it can be without wrinkling the paper it's attached to.

If nothing happens when you touch the battery to the traces, try pressing down harder. If there are foil tape connections between the battery and the wire, try moving the battery so it's touching the traces directly connected to the wire. If the tape connections seem to be at fault, try pressing them down firmly.

Be aware that this project will drain a 9V battery fairly quickly. If the experiment fails, especially if it had been working just a moment before, your battery may be running low. Try a fresh battery.

Step 7: Adaptations and Extensions

Actuated paper is an area of ongoing research. You can experiment with the test platform above by trying out different materials and components, such as:

Try paper that is heavier, lighter, stiffer, or more flexible

Square origami paper, which is lighter than copy paper, is an obvious choice. MIT's Jie Qi built one of her actuated origami cranes using wax paper.

Experiment with different batteries, voltages, and diameters of SMA wire

Instead of a 9V battery, scrounge or buy a battery holder for two or more AA or AAA batteries, and try touching the wires lightly to the aluminum strips. Use the chart on the Dynalloy website (*http://www.dynalloy.com/TechDataWire.php*) to determine the current needed.

Make a more permanent actuated paper model

Instead of masking tape, attach the SMA wire to the paper model by sewing it on with cotton (or other nonconductive) thread. Make a series of small stitches diagonally or perpendicular across the wire. The wire will expand and contract a little more easily and still stay in place.

Play around with the resistance

Since aluminum foil isn't a great conductor of electricity, the longer you make the traces, the more resistance you are adding to the circuit. That, of course, affects the current. Try touching the battery to the traces at different points to see what affect changes in current have on the performance of the SMA wire.

Make more reliable electrical connections

Instead of just overlapping two pieces of aluminum foil tape, try bending the end of the new piece under so that the nonglued side is touching the first piece. Secure it with an extra piece of tape. You can also try thin peel-and-stick copper tape (available from electronics retailers) and solder the connections together. (For directions on how to get started with soldering, see "Project: Make a Souped-Up Solar BEAM Wobblebot" on page 83. For more information on soldering metal tape, see Jie Qi's website tutorials.)

Circuit Basics

There's a lot to know about building electrical circuits, but the basic idea is this: in order for electricity to flow, it has to run in a loop or circle (hence the name *circuit*). Electricity is simply the movement of electrons from one atom to another. An atom, as you probably remember from science class, is one unit of an element, such as oxygen, helium, uranium, or copper. In the center of an atom is a nucleus made up of neutrons and positively charged protons, while negatively charged electrons circle around. A battery or solar cell supplies the force that starts the electrons flowing. For purposes of constructing an electrical circuit for the projects in this book, you can think of electricity as flowing out of the positive terminal (marked with a +) and running back into the negative terminal, or ground (marked with a –). The circuit generally includes a device to use the power being generated such as a light bulb or motor.

If you've never built a circuit from scratch before, you might want to try hooking up an LED bulb or DC (direct current) motor to a battery, just to get the hang of it (but read on before you try this).

You can pick some up from any electronics website or your local RadioShack for a couple of dollars, but you can also rip them out of a dollar store keychain flashlight or old electric toothbrush. As you'll quickly discover, LEDs are fickle. An unused LED usually has a longer leg, or lead, on the positive side, and a flat spot on the negative side of the bulb (an easy way to remember this is that the "plus" side has had something added to it, and the "minus" side has had something subtracted from it). If the leads are not connected to the battery in the right direction—positive lead to positive terminal, negative lead to negative terminal—the LED simply won't light up. The DC motors you will use in this book work either way around, but whether the shaft turns clockwise or counterclockwise depends on which way the battery is facing.

Another thing you'll also need to pay attention to in your circuit is the voltage and the current. Too little and your device won't work. Too much, and you can overheat it. The voltage refers to the amount of energy available in the power source —or more accurately, to the difference in energy levels between two points in the circuit. The current, or amperage (measured in *amperes* or *amps*), indicates how much energy can flow in a certain amount of time. If you think of an electrical circuit as being something like a closed system of pipes filled with water, the voltage is the water pressure, and the current is how fast it flows.

Most LEDs will start to glow with just over 2 volts, and will usually burn out if you try to feed 9V into them. A 3V disc battery like you'd find in a watch should work fine for short bursts, Throwie style (see "Break the Code: Throwies" on page 21). For longer use, you'll have to add a resistor to the circuit. A resistor is a small component that makes it harder for current to flow. In our water pipe analogy, the resistance would be the width of the pipes. The wider the pipe, the more water can flow through at one time. The SparkFun website has an electricity tutorial that gives you a good rule of thumb for using resistors with LEDs. They suggest trying a resistor rated at 330 ohms. If the resistance is too high, the LED won't get as bright as it can, or the resistor will start to get warm. If that happens, just try a lower-rated resistor. If the resistance is too low, the LED will burn out, and you will need to find one with a little higher rating.

Be very careful to avoid making a short circuit by connecting the battery to itself without some sort of device or resistor that will eat up some of the electricity in between. It will quickly overheat and if left too long, may explode or burst.

The other major thing your circuit will need is some kind of wiring to connect the various parts. For fooling-around purposes, you can build a circuit with any insulated wire you have on hand and a little electrical tape. A better tool for prototyping is a supply of short jumper wires with alligator clips on either end that will let you quickly make and take apart connections. For beginners, overlapping the ends of metallic tape is a quick-and-dirty way to make a circuit on paper. But for more reliable connections, soldering is the way to go.

Break the Code: Throwies

The shortest LED circuit you can build is a Throwie. Just pinch a disc battery between the wires of an LED. They're called Throwies because the original version included a small magnet, and were designed to be thrown onto a metal surface such as a bridge, pipe, or fire escape, where they would create an impromptu light-up display. Throwies were invented by Graffiti Research Lab in New York City in 2006, and became an instant real-life viral phenomenon after the group posted a how-to on the website *Instructables.com*.

Electrical Circuit Linkbox

SparkFun Circuit Tutorial (*http://bit.ly/circuit-basic*)

How Circuits Work (*http://bit.ly/circuitwork*)

Science Buddies Electronics Primer (*http://bit.ly/elec-primer*)

Make: Electronics (*http://bit.ly/make-electronics*) by Charles Platt (2009, Maker Media)

Getting Started in Electronics by Forrest M. Mims III (2003, Master Publishing)

Break the Code: Ohm's Law

If you ask an engineer or experienced maker about circuits, don't be surprised if they start quoting Ohm's Law at you. Ohm's Law is an equation that describes the relationship between voltage, current, and resistance. For any given voltage, if the resistance goes up, the current goes down. Ohm's Law says that you can figure out the exact amount of any of the three quantities by using the formula V = I × R, where V stands for the voltage in volts, I represents the current in amps, and R is the resistance in ohms. In other words, the voltage in a circuit equals the current multiplied by the resistance.

If there's not enough resistance in the circuit, you'll have too much current , and too much current can damage electrical components. Ohm's Law is a handy rule to use if you want to figure out how much resistance you need to make your device run without frying it. So if you know the voltage in your circuit and what current the device will take, just plug the numbers into Ohm's Law to get the right resistance.

You won't need to do the math for any of the projects in this book, but if someone tries to explain Ohm's Law to you, you'll be able to smile and nod, because you'll know what they're talking about.

Project: Make an Inflatable Robot

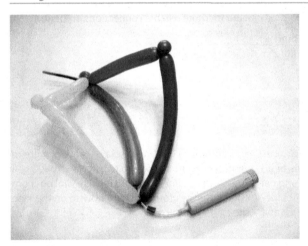

An inflatable robot (or robotic body part) has a soft, stretchy outer surface and one or more air chambers inside. It can move or change shape by adjusting the amount of air it contains.

What It Does

Inflatable robots and robotic parts are cheap, lightweight, strong, and collapsible, making them easy to store or carry. Inflatable actuators, also known as air muscles or pneumatic artificial muscles, allow robots to move in a more natural way than gears and motors.

Where It Came From

If you've seen the Disney hit *Big Hero 6* and its star Baymax, you know that soft, inflatable robots are all the rage. Real-life inflatable robots take their inspiration from the kind of air-filled objects we use every day. In California, inventor (and *Make:* magazine contributor) Saul Griffith and his company Otherlab make air-powered inflatables known as Pneubots (Figure 1-5) that look like elephantine beach toys. Their skin is similar to the material used to make bounce house–type trampolines and the air bladders inside were prototyped using rubber bicycle tubes. Pneubots are soft but strong; the giant-sized Ant-Roach model can easily transport several people on its back. In Massachusetts, iRobot—the same company that makes the Roomba home robovacuum—is working on inflatable military robots that can be folded flat when not in use and carried in a backpack. Baymax itself was inspired by an inflatable robot arm which was developed at Carnegie Mellon University in Pittsburgh to perform caregiving tasks like feeding disabled patients. And at Harvard, researchers are working on silicone robot grippers (Figure 1-6) that close their fingers and multilegged crawlers that can flex and move when air is pumped into specific inflatable chambers.

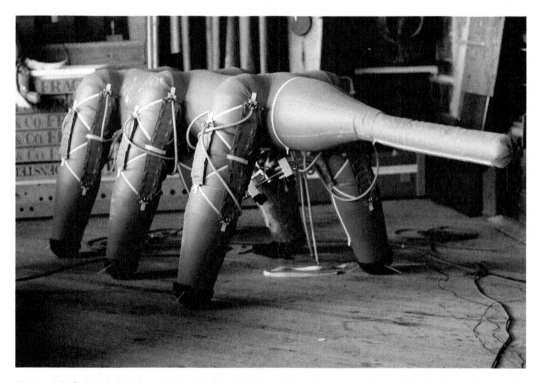

Figure 1-5. *Soft, lightweight, and strong, Pneubotics are inflatable robots. Credit: Pneubotics, an Otherlab company.*

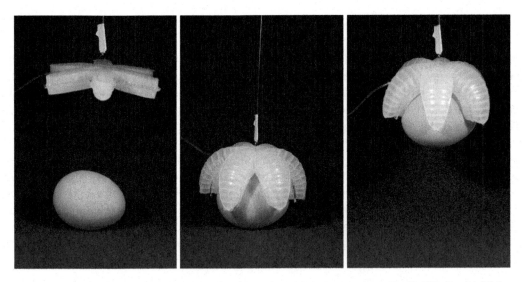

Figure 1-6. *A silicon gripper, activated by inflating it with air, picks up a raw egg. Credit: Filip Ilievski, White-sides Group, Harvard University (photo is CC BY-SA 4.0).*

Air muscles were invented in 1957 by Joseph L. McKibben, a nuclear physicist at the atomic testing site in Los Alamos, New Mexico. McKibben was asked to develop an artificial muscle based on the design of real muscles to help people with weak limbs like his daughter, who suffered from polio. His design used a canister of carbon dioxide gas to expand and contract a rubber tube inside a woven fabric sleeve. The McKibben air muscle is extremely powerful, with the ability to pull 400 times its weight (compare that with a standard battery-powered DC motor, which at best can pull 16 times its weight). Today this style of artificial muscles is used in robotics research as well as for medical purposes.

Most air muscles use rubber tubing, but mechanical engineering student Wyatt Felt created a version using ordinary twisting balloons—the kind used to make balloon animals. Felt's design incorporated an electronically controlled electric air pump that could be programmed to inflate and deflate the robot muscles.

How It Works

Air muscles work by varying the air pressure inside. Like real muscles, they expand and contract in length. An inflatable robot's ability to bend, twist, and stretch has to do with what's on the outside, typically a stretchy material like fabric or rubber. Whether it's hard or compliant depends on how much air or other gas you put into it. At lower pressures, there's less gas inside and the gas molecules can move out of the way more easily when you press on them. The higher the air pressure inside, the more gas molecules you pack in, and the more rigid your inflatable structure becomes.

Think of the tires on a bicycle. Road bike tires, which are designed to get the best mileage on smooth roads, are usually inflated to 130 PSI (pounds per square inch, a unit of pressure). If filled to the correct pressure, a road bike tire is so hard you can't make an indent by pressing on it with your thumb. Mountain bikes are designed to cushion the ride over uneven terrain without puncturing, so they're normally inflated to a softer 30 to 50 PSI. The same principle applies to the Otherlab inflatable robot arm, which can lift as much as a human when the pressure is 200 PSI, but at 20 PSI, can only lift one-fifth the weight.

McKibben-style air muscles have a stretchable inner rubber tube nestled inside a braided mesh fabric sleeve. One end is sealed, and air is pumped in and sucked out through a valve at the other end. The mesh sleeve acts somewhat like a child's Chinese finger puzzle—it gets tighter when stretched lengthwise and loosens when the ends are pushed together. When an air muscle is in the resting position, the mesh sleeve is long and thin, and the rubber tube inside is stretched along its length. Pump air into the rubber tube and it gets fatter. That widens the mesh sleeve and pulls the ends towards each other, making the whole muscle shorter.

The Harvard inflatable gripper, on the other hand, is basically a full-body muscle. Instead of large air chambers, its soft flat body contains narrow air channels. The starfish-shaped gripper is made by pouring liquid silicone rubber into a mold and letting it solidify. The result is a nearly solid but flexible body, filled with a network of air chambers that puff up and pull the body into a curved shape when inflated. Like the claw in an arcade game, the gripper is lowered over an object. Pump in some air, and its fingers curve down and grab whatever is below.

Wyatt Felt's twisty balloon air muscle uses yet another way to move. It differs from the McKibben muscle in that it pushes instead of pulls. When deflated, Felt's balloon air chamber is short and narrow. When activated, it expands in both length and width. Felt's display model used an electronically controlled air pump with intake and outflow valves. The muscles were hooked up to "legs" made of curved pieces of plastic that straightened out when actuated by the muscle. But on his first prototype, both the leg structure and the air muscle were made from balloons, and air was supplied by a hand pump.

Making the Project

This project is based on the prototype twisting balloon air muscle created by Wyatt Felt, and adds a few twists of its own. Instead of legs that curve and straighten out, this version opens and closes a *Sarrus linkage*, a pair of hinges set at right angles to one another, similar to the ones used in pop-up paper robots like the RoACH. This adaptation of Felt's air muscle also incorporates a built-in release valve mechanism that takes only a minute to create.

Project Parameters

- Time Needed: 1 hour
- Cost: Less than $10
- Difficulty: Easy
- Safety Issues: Watch out for popping balloons!

What You Need to Know

- Skills You Already Have: Blowing up balloons
- Skills You Will Learn: Balloon twisting techniques

Gather Your Materials

- Twisting balloons (available in party goods stores
- Balloon hand pump
- Flexible vinyl, 1/4-inch (15 cm) diameter (available in hardware stores)
- Electrical tape
- Scissors

Directions
Step 1: List Your Requirements

The goal of this project is to make a soft, compliant robotic part using inflatable rubber materials for both body and actuator.

Step 2: Plan Your Project

This build doesn't require a lot of planning, unless you want to get fancy and try one of the extensions below. Just make sure you have plenty of balloons on hand, since they are not as robust as other kinds of inflatable construction.

Step 3: Stop, Review, and Get Feedback

You will probably not be surprised to learn that people like MIT's Erik Demaine have studied the math behind balloon twisting. You can learn more about it in papers by Demaine and "recreational mathemusician" Vi Hart (see "Inflatable Robot Linkbox" on page 31).

Step 4: Build Your Prototype

Follow these steps to build the muscle balloon:

1. Use the hand pump to inflate two balloons, leaving about 4 to 5 inches (10 to 12 cm) uninflated at the end. This is known as the tip of the balloon. Remove the pump and let out a little air (known in the business as burping the balloon). This bit of slack makes it easier to twist the balloon without popping it. Tie the neck of the balloon in a knot to seal it. Helpful hint: before inflating a balloon, stretch it lengthwise a few times.

2. Take one of the balloons and pinch it gently about 3 inches (8 cm) from the knot. Twist it around three times. Do the same to the other balloon, then connect the two balloons by twisting them together where they are already twisted.

3. To make a hinge in the first balloon, bend it in half. At the bend, pinch a golf-ball-sized segment in your fingers.

Twist it around three times, spinning it like a dial. Then circle it around the balloon itself until it reaches its starting point. The hinge should look like a knee sticking out in front. Do the same with the other balloon.

4. Tie the tips of the balloons together using the uninflated extra rubber. Tie another knot about half an inch (1 cm) above the first. The balloons will form a diamond shape with the hinges in the middle.

5. Now take the vinyl tubing and cut a piece 10 inches (25 cm) long. Make a release valve in the piece of tubing by taking the scissors or an art knife and cut a small slit about halfway down. Don't let it go more than partway through the tubing. You should be able to bend the tubing back to open the slit without tearing the tubing. Take a piece of electrical tape about 2 inches (5 cm) long, and fold over a tiny bit at one end. Take the other end and wrap it around the tubing so it covers the slit. Use the folded-over end as a tab that you can pull back to reveal the slit. Try bending the tubing back so the slit opens up. Then reseal it with the tape. You can test it by inserting the air pump to make sure it is airtight when sealed.

6. Poke one end of the tubing through the gap between the two knots in Step 3. Take a third balloon, inflate it, then let the air out. This is your air muscle. Pull the opening of the air muscle balloon over the end of the tubing that pokes out between the knots. The balloon should cover about 1 inch (2 cm) of the tubing. Secure the balloon to the tubing with a piece of electrical tape.

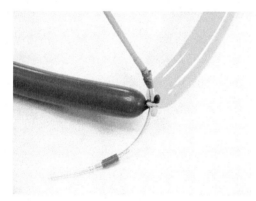

7. Take the top and the bottom of the balloon diamond and press them towards each other. This is the movement your inflatable hinge will make. Decide how close you would like them to get, and tie the tip of the air muscle balloon to the tips of the other two balloons to hold them in this position.

8. Insert the end of the air pump into the other end of the clear tubing, as far as it will go. Secure it with more electrical tape if needed.

Step 5: Test Your Design

Use the pump to slowly and carefully inflate the muscle balloon. As it fills with air, it should lengthen and push the balloon hinge open. To let the hinge close up again, open the release valve by unwinding the tape enough to explore the slit, and bending the tubing back to widen the opening. The air should escape and the balloon should return to roughly the same length as when it started.

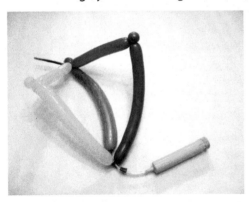

Step 6: Troubleshoot and Refine

If you're having trouble inflating the balloon, test out your pump on another balloon fresh out of the package. Cheap pumps break easily. Also check your balloon for leaks.

Step 7: Adaptations and Extensions

Expand on the balloon body

Add a complete robot to the robotic body part with a torso, arms, and a head. (See "Inflatable Robot Linkbox" on page 31 for a tutorial.) Or go abstract with a mathematical balloon model.

Make a splitter to actuate several robotic parts at once

Take a short piece of clear tubing and cut out a diamond shape big enough to fit another piece of tubing. Make as many holes as you want extra air lines. Insert another piece of clear tubing into each juncture. Seal around it with Sugru or with a hot glue gun to prevent leaks.

Try turning your balloon into a McKibben muscle

Insert a twisty balloon into a Chinese finger puzzle or short length of braided expandable sleeve (available as an electronics cable organizer or a fidget toy). A YouTube tutorial will show you how (see "Inflatable Robot Linkbox" on page 31).

Make a Harvard silicone gripper with a 3D printed mold

Read about 3D printing in Chapter 2, then check out the Instructable from Harvard-trained roboticist Ben Finio to make your own mold for creating an air-actuated silicone gripper.

Inflatable Robot Linkbox

Wyatt Felt Twisty Balloon Pneumatic Actuator (*http://bit.ly/twisty-balloon*)	Dylan Gelinas simple balloon robot tutorial (*http://bit.ly/gelinas-robot*)
Vi Hart on balloon math (PDF) (*http://bit.ly/math-balloon*)	Phil Teare braided sleeve balloon McKibben muscle tutorial (*http://youtu.be/cc5Ge49eL2A*)
Basic balloon twists (*http://www.professorwonder.com/twists.htm*)	Harvard silicone gripper Instructable (*http://bit.ly/robo-gripper*)

Robots That Get Around

2

How often I found where I should be going only by setting out for somewhere else.

— R. Buckminster Fuller

One of the most interesting challenges in robot design is figuring out how to get from place to place. Traditionally, robot locomotion has fallen into two categories: bipeds that walk in something approximating a human gait, and rolling robots that drive around like vehicles. But in recent years the options for robot locomotion have taken off. Today it's not unusual to see robots that can fly, swim, climb, creep, or crawl—and some that can move in ways that are hard to categorize. The advantage of these alternative modes of transportation is that they allow robots to access places humans can't easily go, from collapsed buildings to other planets.

From a design point of view, robots that move in unexpected ways are also fun to play with and fascinating to watch. Just look at the popularity of tiny drone copters, which can be remote controlled or preprogrammed with a flight pattern, or which carry onboard sensors that let them avoid obstacles on their own. The ability to shrink down electronic flight components to the size of a deck of cards has created a boom in the DIY arena as well. Hobbyists can assemble their own onboard stabilizer, compass, gyroscope, accelerometer (a sensor that measures changes in speed and direction), and GPS, for just a few hundred dollars and attach them to the body of their choice. Brooklyn Aerodrome sells a kit on the Maker Shed website that fits the flight deck on a corrugated plastic board that can quickly be switched to a new construction foam body after heavy use. One standout exhibit at the Emma Willard Mini Maker Faire in Troy, New York in 2013 was the flying pizza box put together by 10-year-old Emma Edgar, with help from her dad, Marc. The flattened large-size pizza box can easily maintain heights of 100 feet or more for more than 30 minutes and needs minimal human input to stay aloft.

Even rolling robots can move in unconventional ways. The Segway personal transporter is actually a robotic platform that can balance itself on a single axle. Designed by inventor (and FIRST Robotics Competition founder) Dean Kamen, the Segway was designed to carry a human being, but it is also used as a base for other kinds of mobile robots. Or consider the Sphero Robotic

Ball (Figure 2-1). Controlled by a smartphone, the Sphero can self-propel across a room by rolling, and play games with both humans and pets. Its waterproof casing contains a gyroscope, accelerometer, and LED lights.

Figure 2-1. *Sphero is a robotic ball that can be programmed to jump or roll with a smartphone. Credit: Orbotix.*

Researchers looking for new ways to make robots move often turn to the animal kingdom. MIT's RoboTuna and other robotic fish that swim like their real-life counterparts have been around for a long time. Harvard's RoboBee is smaller than a paperclip and can flap its wings, hover, and turn on command. Still in the development phase—they've yet to come up with a battery small enough for the RoboBee to carry—the microdrone might someday help with environmental monitoring, or crop pollination. Boston Dynamics, which became a part of Google in 2013, is famous for Big Dog, its aptly named experimental rough terrain quadruped. Resembling a headless Doberman on steroids, Big Dog can trot along through rubble, mud, snow, and shallow water carrying more than 300 pounds on its back. The company's fastest robot to date is the Cheetah, which can reach 30 miles an hour on a treadmill. But Boston Dynamics is also working on RiSE, a compact, six-legged robot that can scale brick walls and rough treetrunks at a much more leisurely pace. Inspired by the gecko lizard, RiSE uses microspines on its toepads to latch onto surfaces, balancing with the help of its flat triangular tail.

A lot of the cutting-edge research into robot locomotion has been supported by the U.S. military through DARPA, the Defense Advanced Research Projects Agency. Driverless cars, robotic drone aircraft, and remote-controlled miniature tanks used in battle zones and rescue situations are just part of DARPA's legacy. A series of design competitions known as the DARPA Robotics Challenge has also produced some of the most mind-blowingly strange modes of transport in the robot world—among them undulating robo-worms and amoeba-like machines able to ooze and creep along thanks to flexible materials.

The projects in this chapter will give you the chance to build two different systems of robot locomotion: the tensegrity robot, which uses tension, compression, and vibration as modes of transportation; and the wheel-leg, a hybrid that combines some of the best aspects of both types of traditional robot motion.

Locomotion Linkbox

Brooklyn Aerodrome (*http://www.brooklynaerodrome.com/*)

Segway Personal Transporter (*http://www.segway.com*)

Sphero (*http://www.gosphero.com/*)

Harvard RoboBees (*http://robobees.seas.harvard.edu/*)

Boston Dynamics (*http://www.bostondynamics.com/*)

RiSE (*http://kodlab.seas.upenn.edu/RiSE/Home*)

DARPA Robotics Challenge (*http://www.theroboticschallenge.org/*)

Project: Make a Compressible Tensegrity Robot

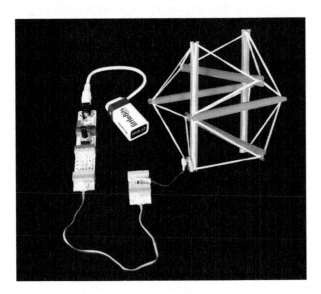

Figure 2-2. *Finished tensegrity robot*

A tensegrity robot is made up of rigid struts suspended by a web of cords under tension, which let it move and respond to changes in its surroundings.

What It Does

A tensegrity structure can flex, stretch, compress when dropped or pressed, and then spring back into shape. It also has a high degree of compliance, a robotics term that means it won't harm people or equipment around it. That, together with its resilience, makes the tensegrity a useful framework for robots that need to withstand jolts or squeeze and twist themselves through irregular spaces. But it also offers a range of unique ways to mobilize a robot, including compression and vibration of its flexible cords.

Where It Came From

The word "tensegrity"—a portmanteau combining "tension" and "integrity"—was coined by renegade architect Buckminster Fuller. Many of Fuller's inventions, including the geodesic dome, play around with tension and compression in their structural design. Although Fuller is the person most closely associated with the concept, most scholars agree that he was inspired by artist Kenneth Snelson, whose sculptures incorporate what he prefers to call "floating compression." Snelson's massive sculptures do seem to defy gravity as they spread and soar skyward. His most famous work is "Needle Tower," a 60-foot-high tapering sculpture of aluminum tubes and steel cables on display at the Hirshhorn Museum and Sculpture Garden of the Smithsonian Institution in Washington, DC.

Roboticists are looking at ways to employ tensegrity in robot design. At the NASA Ames Research Center in Mountain View, California, scientists Adrian Agogino and Vytas SunSpiral were playing with a squishable stick-and-elastic baby toy when they got the idea to use the tensegrity's forgiving structure (Figure 2-3) in the design for a mission to Saturn's moon Titan. Unlike the 2005 Cassini spacecraft's Huygens lander, which scientists believe sank below Titan's crust upon landing, their Super Ball Bot lander could bounce safely and tumble along Titan's surface. And the Super Ball Bot could be dropped from the spacecraft without the need for an elaborate skyhook mechanism like that used by the latest Mars Curiosity rover. They are now studying how to protect the instruments a tensegrity lander would need to carry.

Other possible uses of tensegrities include TetraSpine, a multisegmented tensegrity robot being developed at Case Western Reserve University in Cleveland, which can crawl over uneven surfaces. At Union College in Schenectady, computer scientist John Rieffel and his students are working on using vibration frequencies to steer ball-shaped tensegrities (Figure 2-4). And Erkin Bayirli, an architecture student in Vienna, has been building elegant tensegrity robots out of bamboo that can walk or crawl by pivoting on their corners, without legs.

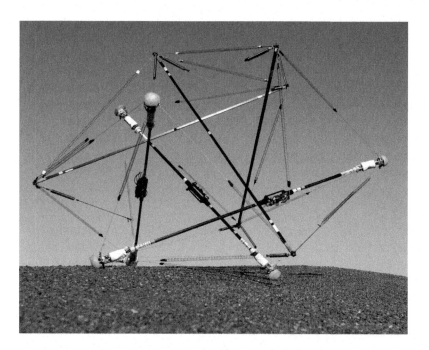

Figure 2-3. *A prototype of NASA's Super Ball Bot built by Ghent University's Ken Caluwaerts.*

Figure 2-4. A tensegrity robot developed at Union College which can be steered using vibration. Credit: Steve Stangle.

How It Works

A tensegrity structure maintains its shape by balancing the push and pull of its various components. In a standard tensegrity, cords do not touch cords, and struts do not touch struts. Each cord is stretched between two rigid struts. Each strut has cords at either end pressing inward upon it. The tensegrity transfers these forces throughout the entire structure, giving it exceptional strength and resilience. For example, a bicycle wheel is a kind of tensegrity. The thin metal rim is kept perfectly round, or "true," because it is being pulled evenly by all the spokes that connect the rim to the hub in the center. That gives it as much strength as a wagon wheel, without the need for heavy wooden spokes. But the tension and compression in a tensegrity must be in balance or it will pop and twist like a bicycle wheel with a broken spoke.

The complex interplay of forces in tensegrities make them intriguing to mathematicians and physicists. And they may also be useful in explaining how bones and connective tissue work together in living things. "Biotensegrity" is a term used by orthopedic surgeon Stephen Levin to describe a model of the skeleton that he believes is more accurate than the standard way of looking at it. In the biotensegrity model of the human skeleton, the struts are the bones, and the cords are the web of tendons and ligament that hold them together. As Levin points out, the bones in a body are not connected to each other directly by hinges, like a door swinging on a doorframe. Instead, the bones float in a network of connective tissue called the fascia. They slip past each other as they move, like the parts of a tensegrity sculpture. The fascia help to distribute the force applied on the bones by the muscles throughout the entire system, resulting in a body that is much more resilient to stress than it would be without it. With this in mind, roboticists are now trying to use the biotensegrity model to make robot bodies more springy and lifelike.

One interesting quality of a tensegrity is the way it can be pushed flat and then snap back into shape. So a robotic tensegrity can move by bouncing, like the Super Ball Bot. Or it can be compressed and then expand in waves. Robots like TetraSpine are made up of a column of self-contained segments connected to each other by cords. Each segment can compress in relation to its neighbors, like discs in a human spine. To move, TetraSpine uses DC motors and an Arduino microcontroller board to tighten and loosen the elastic connectors, causing waves that propel the robot along. The waves are created by central pattern generators, similar to the rhythmic signals that help living vertebrates crawl, walk, swim, and fly automatically, without input from the brain. A virtual simulation of TetraSpine has been able to use its rhythmic motion to traverse several kinds of irregular terrain.

Because the cords in a tensegrity are stretched tight, they can also vibrate like the string on a guitar. The Union College tensegrity robots move by setting up vibrations along the elastic cords. This is done by attaching one or more vibrating motors to one of the struts, making the strings resonate. Different speeds create different wave patterns. These patterns can be used to make the tensegrity jiggle and shake in one direction or another. Yet another way of moving is that used by Erkin Bayirli's bamboo tensegrity robot, Shi. On the top of the robot is a servo motor that oscillates back and forth. As it moves, it twists a strut at the rear from side to side. Like the stick controlling a marionette's legs, the moving strut pulls first one corner and then the other slowly forward.

Perhaps the most exciting aspect of research into tensegrity robots is Rieffel's idea that mechanisms can act as minds. He is exploring whether control normally associated with a robot's "brain" can instead be outsourced directly into body dynamics. In other words, the structure of the robot would have its own intelligence. When stress from the outside is applied, the robot structure reacts in a way preprogrammed by its physical design. As Rieffel explains, this mechanical programming can free up a robot's computational resources for higher-level tasks, like tracking objects or detecting survivors trapped in rubble. The idea of programming through physical structure takes the importance of robot body design to a new level.

Making the Project

To understand how tensegrities fit together, you have to build one yourself. But as Rieffel notes, assembling a tensegrity can be challenging because of what is called "pre-stress stability"—when you move one part, everything else moves, too. So to start off, you'll make a version that's flimsier but easier to handle. The directions for assembling this six-strut tensegrity out of drinking straws and rubber bands are based on a DIY tensegrity holiday ornament project from Bre Pettis that appeared on the Make: website in 2007. It cleverly eliminates the need for an extra pair of hands by using small rubber bands to hold the straws in place until you have assembled the entire tensegrity. Then the support rubber bands are cut away and the structure pops open into its final form.

You will be building this project in two steps: the tensegrity body, and the electronic circuit. With such a lightweight structure as your base, you'll be able to take advantage of another timesaver—the littleBits line of electronic building parts. Once you've built your tensegrity, you can quickly put together a circuit to make your robot move. The circuit will consist of a tiny vibrating motor, a dimmer switch to make it run faster or slower, and a bar graph indicator that shows roughly how much power you're supplying to the motor. Varying the speed and placement of the motor will produce different kinds of motion, giving it a kind of physical intelligence.

 Don't forget to document your work!

Project Parameters

- Time Needed: 1 hour
- Cost: $75–$100 (for reusable littleBits modules, purchased separately or in a kit)
- Difficulty: Easy to moderate
- Safety Issues: None

What You Need to Know

- Skills You Already Have: Cutting, following directions
- Skills You Will Learn: Assembling a circuit

Gather Your Materials

- Six or more drinking straws (it's good to have extras)
- Six rubber bands, roughly 5 inches long
- Six additional rubber bands (preferably shorter)
- Scissors
- Masking tape, glue dots, double-sided mounting tape or other removable adhesive
- littleBits modules
  — 9V power supply
  — Dimmer switch
  — Bar graph (LEDs)
  — Wire(s)
  — Vibrating motor

Directions

Step 1: List Your Requirements

The goal of the tensegrity robot project is to build a self-propelled object based on the push-and-pull design of a tensegrity structure. Its movement should be controllable by altering the arrangement of its parts and the speed of its motor.

Step 2: Plan Your Project

Following the drinking straws-and-rubber band design outlined here is the most reliable way to get started successfully building tensegrities. The area most open to improvisation at this stage is the design and placement of the littleBits circuit. In particular, think about how to attach the motor so that it interferes with the motion of the tensegrity structure as little as possible. You can list or sketch out different possibilities to try once you have everything assembled.

Step 3: Stop, Review, and Get Feedback

Some helpful tips for building your first tensegrity are to take your time and refer back to the pictures often. If you're not sure what you're doing, just go with the flow. It may not be clear how it works until you reach the last step.

Keep some spare straws on hand while you're working. If a straw bends, you're better off replacing it than trying to fix it.

Step 4: Build Your Prototype

1. Cut six pieces of straw to no more than about 5 inches.

2. On each straw, cut a slit on either end, making sure that the slits are aligned (i.e., both vertical). The slits should be around a quarter of an inch deep—enough to hold the rubber band in place, but not so much that the straw begins to weaken and bend.

3. Line up two straws and wrap a small rubber band loosely around each end of the pair. Do the same to a second pair of straws and slide them perpendicularly between the first two straws to form an "X."

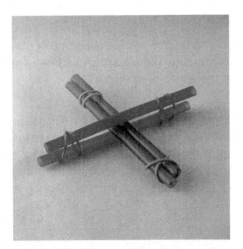

4. Take the last two straws and wrap a small rubber band around one end. Slide them through the intersection of the other straws so that they are perpendicular to the first two pairs. When they are in place, wrap a small rubber band around the other end.

5. Add the long rubber bands. Line up the slits in one pair of straws so that they are horizontal and hold them so that the ends are facing you, one above the other. Fit a long rubber band into the slit of the upper straw end facing you. Hold the remainder of the rubber band up and stretch it over the ends of the pair sticking up, fitting it into the slits at the top of the straws. Then stretch the rest of the rubber down and latch it onto the horizontal slit in the other end of the first straw. The result should look like a suspension bridge.

6. Do the same with the all the remaining straws. Adjust the rubber bands so they are even.

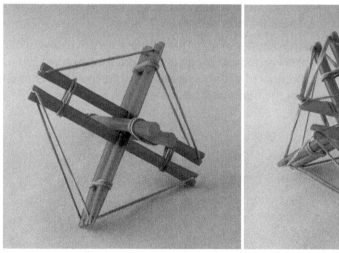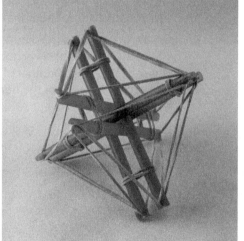

7. Cut away the small rubber bands so that the tensegrity springs open. Make sure none of the straws are touching. If needed, adjust pairs of straws again so they are parallel.

Now it's time to assemble the littleBits electronic circuit that will make your tensegrity go.

1. Plug the power piece into the 9V battery.

2. Next, attach the dimmer switch. Technically, this component is called a *potentiometer*. It lets you turn the voltage from the battery up or down.

3. Connect the bar graph to the dimmer switch. This is a piece with five rows of miniature LED light bulbs. The more power going through the bar graph, the more LEDs will light up. It doesn't tell you the number of volts going through, but you can record different voltage readings in terms of the number of LEDs that are lit up.

4. Attach one or more wires. The wire modules are short, so use two or three if you have them. These give your tensegrity robot some space to move without having to drag around the rest of the components.

5. Finally, add the vibrating motor. This is a small disc, about the size of a vitamin pill, with two thin wires attaching it to a base that snaps onto the wire. Be careful not to break the wires or tear them out of the component.

Building Robots with littleBits

The littleBits electronic building modules are one of the hottest Maker tools around. Invented by MIT Media Lab alum Ayah Bdeir, littleBits are an ever-growing library of postage-stamp-sized components that snap together magnetically. All the circuitry needed to make the components work together is already built in, and they are all open source. (*Open source* means a design is made public for anyone to use, improve, and distribute to others.) The modules come in sets, but you can also buy them individually. Originally started as an educa-

tion initiative to help children and other beginners design simple circuits and electronic projects, littleBits are being used by artists, designers, and inventors to aid in rapid prototyping.

What makes littleBits different from other electronic toys for kids is their stripped-down appearance. That has good points and bad points. On the plus side, they're small and light, so they fit into tight places. All the components and connections are exposed like a circuit board, so you can see how they're put together. Tiny labels help you figure out what each component does, and some modules can be adjusted with slider switches or a tiny screwdriver (included). But unlike Lego or other construction toys, littleBits are not themselves building blocks. In order to make something with them, you have to attach them to something made out of materials you provide. And because there's no casing to protect the circuit boards, they're a little less rugged than traditional electronic building sets.

That said, getting started with littleBits is easy. Each piece attaches magnetically to the next—just click them together and you're done. If the magnets try to repel each other instead of attracting, you know you've got a piece the wrong way round. Creating a circuit is also a simple process, because of the color coding. Blue pieces are for power—9V or coin battery or USB plug. Pink is for input, such as switches, sliders, buttons, light sensors, touch sensors, etc. Orange pieces are connectors—wires and logic gates. And the green pieces are the output devices, including LEDs, motors, and speakers of various kinds.

The littleBits modules and kits are on the expensive side compared to many kids' building sets and to raw components you need to wire up yourself. (The basic starter kit costs about $100, with individual modules ranging from around $8 to $40.) And once you start figuring out what you can do with the pieces, you'll quickly find yourself in need of extra modules, since the sets don't contain multiples of the same part. It pays to design your project carefully to make the most of the littleBits you've got on hand. When attaching modules to a project, think about how to make it easier to reuse the parts later. Here are some ways to attach littleBits parts to your project so they can be removed later with minimal harm to the piece or your model:

- Masking tape—take a piece about 3-4 inches long and roll into a loop, sticky side out, fastening it to itself

- Adhesive dots (such as those sold under the brand name Glue Dots)

- Modeling dough (homemade or Play-Doh)

- Velcro straps

- Clear mounting tape, such as 3M VHB or Scotch Permanent Clear Mounting Tape

Step 5: Test Your Design

To try out your tensegrity robot, attach your electronic circuit to your straw model. You'll want to situate the vibrating motor so that none of the littleBits modules get in the way of the tensegrity structure's motion around the table.

1. Decide where you'd like to attach the disc end of your motor. Use tape or another adhesive (see "Building Robots with littleBits" on page 45) to hold it onto one of the struts. Play out the motor wire along the strut. Attach the motor base and the wire base (which you have connected magnetically) to the same strut.

2. Thread the orange wire modules through the tensegrity to keep the remaining components out of the way. You might want to try holding them above the structure so they don't touch any of the other struts or the table.

3. Turn on the vibrating motor. Slowly increase the power to it using the dimmer switch. You should start to see the rubber bands vibrate in sympathy. Notice how different speeds produce different movements. At a certain point your tensegrity should start to shimmy along the table. See if you can steer it right and left by turning the power up and down.

4. Experiment with placing the motor in different locations on the tensegrity—in the center, off on one corner—to see which position produces the most reliable and interesting movements.

Step 6: Troubleshoot and Refine

If your tensegrity doesn't move, make sure that it isn't caught up on any of the components. You can also try putting the weight of the components higher up or lower down on the tensegrity. You may need to move its center of gravity a little off-center to overcome its inertia.

Step 7: Adaptations and Extensions

You can build a more permanent device using materials that are sturdier but take more effort to prepare, such as wooden dowels. Other suggestions:

- Elastic cord
- Fishing line
- Clear plastic beading thread
- Pencils
- Wooden dowels
- Bamboo garden stakes
- PVC pipes
- Eye screws
- Plastic end caps

To propel it, you may have to use a more robust motor, such as the littleBits DC motor. Convert it into a vibrating motor by taping a small weight (such as a bead or metal nut) onto the motor's shaft. You can also add additional littleBits parts—including the remote control modules, the number module with a digital readout, or even the programmable Arduino module—to make either the prototype or more permanent tensegrities more interactive. For a real challenge, do some research on tensegrities and tensegrity robots to find more advanced models to try to animate.

Project: Design a Wheel-Leg Hybrid

A wheel-leg hybrid is just what it sounds like—a blending of the basic wheel/axle design with one or more legs.

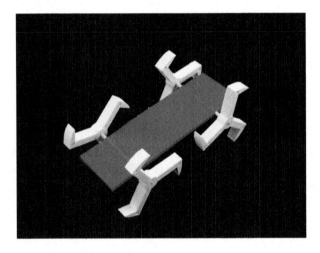

Figure 2-5. *The finished wheel-leg robot*

What It Does

The wheel-leg makes it possible for robots to benefit from the efficiency of wheels on smooth surfaces while using its legs to keep going over rougher terrain.

Where It Came From

There are several variations on the wheel-leg design in the world of robotics. One design, the Quattroped, is a four-wheeled/legged robot developed at the Bio-Inspired Robotic Laboratory at National Taiwan University. In a move worthy of Optimus Prime, it can transform its wheels into legs and shift from rolling to walking in under a minute. Each wheel folds itself in half vertically into a "C" shape to become a leg. The central hub then slides up to become a "hip joint." The legs can rotate 360 degrees around and grab the terrain with their rubber treads. Although not particularly graceful, the Quattroped can pull itself up stairs and over tree roots, and boogie on the laboratory's smooth floors.

An even simpler design is the Wheg from Case Western Reserve University in Cleveland. A Wheg —another portmanteau word that blends "wheel" and "leg"—works a little like a wagon wheel with the rim removed. Each spoke functions as a leg, but the whole thing revolves around a central axle, just like a traditional wheel. More elaborate versions of the Wheg have been developed, such as the Loper from the University of Minnesota, which has a rounded lobe at the end of each of its three legs, allowing it to roll a little with each step. The MSRox wheel-leg designed from the University of Iran actually has a separate wheel on the end of each spoke. But the basic Wheg design by itself is simple and effective.

Break the Code: Triskelion

The concept of a wheel made of legs goes back much farther than the electronic age. It appears in the ancient Celtic symbol known as a *triskelion*, which is Greek for "three legged." In most depictions, the legs of the triskelion appear to be joined literally at the hip as they gallop around the circle. A bare-legged version with a leaf-framed face in the center still appears on the flag of Sicily. Another interpretation, legs clad in medieval armor, graces the flag of the Isle of Man.

How It Works

The invention of the wheel is considered one of humanity's greatest achievements for a reason. The mechanics and control involved in rolling a cart downhill are so much simpler than those required to run after it. Given a relatively even surface, a vehicle on wheels can usually move faster and more smoothly than even the fastest sprinter.

That said, if you're a land animal, legs are the way to go. Legs have an advantage when it comes to climbing over obstacles. But getting robots to walk like people or animals isn't easy. They need programming and sensory feedback to tell them which leg to move when, respond to differences in terrain, and stay upright while doing it. That's why a design that combines the two can add up to more than the sum of its parts. On smooth surfaces, the wheel-leg can take advantage of the efficiency of a standard wheel. Off-road, it can handle changing surfaces more easily and climb over obstacles.

For robots, the Wheg-type wheel-leg has another advantage over the traditional hip-socket type limb: it can spin around an axle like an ordinary wheel, so it doesn't need programming to help it with balance and alternating stride.

Wheel-Leg Linkbox

National Taiwan University Quattroped (*http:// bit.ly/cvWKr0*)

Case Western Reserve University Whegs (*http:// biorobots.case.edu/projects/whegs/*)

Loper from the University of Minnesota (*http:// distrob.cs.umn.edu/loper.php*)

MSRox (*http://bit.ly/climbing-robo*)

How Does a 3D Printer Work?

Three-dimensional or 3D printing (see Figure 2-6) has transformed the way researchers and hobbyists make prototypes. It's now possible to create, test, and tweak a new design in minutes or hours instead of days or weeks. Even more extraordinary is how quickly 3D printers have gone from an industrial tool costing thousands of dollars to something tinkerers, artists, and small startup businesses can buy or make themselves for just a few hundred dollars. Companies like MakerBot and open source projects like RepRap are bringing 3D printing to homes, offices, and makerspaces around the world. And for those who don't have access to a machine themselves, online services like Shapeways, Ponoko, or Sculpteo make it possible to upload a file (or choose a design on the website) and have it printed and mailed to your home.

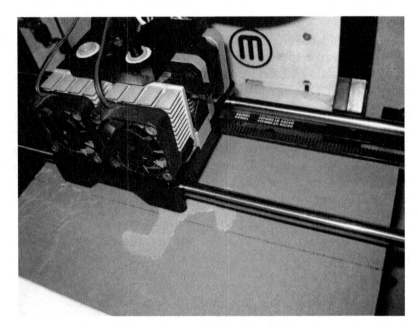

Figure 2-6. *A 3D printer producing the wheel legs from this section*

Robot designers use 3D printers to create custom parts as well as entire bodies. Bob the BiPed is a 3D-printed robot created by hobbyist Kevin Biagini. Aside from a boxy body and two short stubby legs, the only other parts needed to build Bob are a small microcontroller board for a brain, a couple of ultrasonic sonar sensors for the eyes, and two servo motors to move the legs. Directions for making Bob are available on Instructables. More ambitious is Jimmy, an open source cartoon-like humanoid robot designed by Intel Director of Future Casting Brian David Johnson. As part of the 21st Century Robot project, anyone will be able to access the files to print their own Jimmy body and modify his design as desired. Plans call for a kit users can buy to animate their own Jimmy.

But the most exciting use of 3D printing in robots so far has to be the Robohand, a prosthetic developed by South African carpenter Richard Van As. He started working on creating his own artificial hand after losing four fingers in a work accident. Using donated MakerBots, Van As and Ivan Owen, an American collaborator, came up with a design that was cheaper and more functional than experimental models available commercially. They released the plans for free on Thingiverse, a file-sharing site for users of MakerBot. Then, in 2014, Jon Schull of MAGIC, a collaborative innovation center at the Rochester Institute of Technology, founded a group called e-NABLE that matched volunteers with 3D printers and children with missing hands or fingers to create custom-fitted Robohands at a fraction of the cost of traditional prosthetics.

The standard 3D printer used at home or the neighborhood makerspace works by taking a spool of *thermoplastic filament*—a plastic thread that becomes soft enough to flow when heated—and feeding it through a movable printer head, or extruder. The computer-controlled extruder squirts out a thin stream of softened plastic, building it up layer by layer to create the shape you have programmed in. It's akin to squeezing frosting out of a piping tube—in fact, one of the first home 3D printers was called the Cupcake CNC. (*CNC* stands for "computer numerical control," a term also used to describe computer-controlled routers that can be programmed to cut out patterns from wood and other materials.)

The two most common types of plastic used in 3D printers are ABS (the same plastic used to make Lego bricks) and PLA (a biodegradable plastic made from cornstarch, tapioca, or other plant material). You can also get filament made from a wood-plastic mixture, and more varieties are being developed all the time. The Pancake Bot, a 3D printer made using a Lego Mindstorms Robotics kit, was a big hit at World Maker Faire New York in 2012 (see Figure 2-7). Its computer-controlled ketchup-bottle extruders squeezed batter onto a griddle platform, creating tasty sculpted hotcakes. Other materials like plastic, glass, ceramic, and metal can also be 3D-printed using a method called *sintering*. This technique involves aiming lasers at a bed of powdered material to fuse it together into a specific shape with incredible detail.

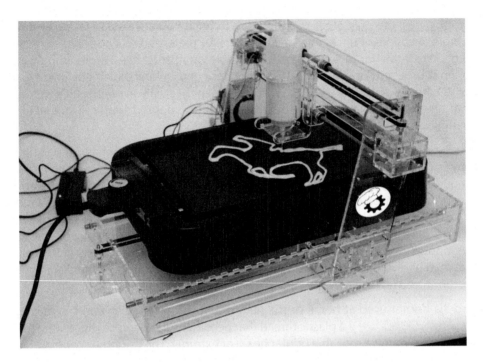

Figure 2-7. *The Pancake Bot in action*

With schools, offices, and copy places beginning to get their own 3D printers, it may seem like 3D printing should be as easy as making a copy on a Xerox machine. Not quite. There's still a lot of preparation and cleanup needed to produce a successful print. And just like the early days of the Xerox machine, 3D printers still require a fair amount of trouble-shooting. When something goes wrong, you may have to check the temperature of the printer platform, how fast the filament is being fed to the extruder, and on and on.

Before you can even print a new design, it's important to be sure it works within the limitations of the machine you're using. Things to think about include:

Does the object have a good base?
Since the printer builds objects from the bottom up, you need to make sure it adheres to the printer platform. The printer software can help you with this by generating a removable "raft," a thin base of plastic underneath your object.

Is the orientation optimal?
Each new layer of your object has to be supported by the layers below. Holes or gaps can be a problem, but sometimes just flipping an object on its side will make it possible for a design to print successfully. When parts of the printed object stick out enough to sag before the plastic has hardened, the printer software can generate a lattice of support material to hold it up. On printers with two or more extruder heads, you can set up the machine to print the support in a different material, making cleanup easier.

How much time and material will it take?

You can adjust the resolution on the printer, just like you can on a 2D machine. Lower resolution prints have less detail, but take less time, too. Depending on the design, you can also save time and material by choosing to make the object hollow or honeycombed inside instead of solid. In addition, you can specify how many *skins* (aka *shells*) to print, which tells the machine how thick to make the outer walls.

Your print may be a little rough when it's done, with ridges where the layers were laid down. These bits of excess material, along with the raft and any support material, can be snapped off by hand and smoothed with a nail file or sandpaper. A more advanced method is to give your printed object a nice shiny finish by exposing it to acetone (nail polish remover), a solvent that dissolves ABS plastic. This has to be done carefully and with good ventilation, because the fumes are both flammable and toxic.

Although they're not quite foolproof enough for the average person, it's clear that 3D printers are quickly becoming the go-to tool for creating instant models, one-off parts, and unique toys and artwork. It won't be long before they take their place in most households, alongside the paper printer/copier/scanner and other devices that are now a part of our lives.

Making the Project

The Wheg type wheel-leg makes a perfect subject for practicing prototyping techniques. It can easily be built in one piece, yet it offers an infinite number of variations to try out. And it provides a great opportunity to learn how to use 3D modeling software and 3D printers. You don't even need to construct an entire robot to test it out. Just make a rolling platform, like a child's pull-toy, or hack an old RC car by replacing the standard wheels with wheel-legs, and see how your design performs in different environments.

To get you started with CAD (Computer Aided Design) software, directions for re-creating a basic three-wheeled wheel-leg are outlined below. It uses a free online program called Tinkercad, so you don't even need to download any software to your computer.

 Don't forget to document your work!

Project Parameters

- Time Needed: Several hours or more
- Cost: Roughly $10 per piece for plastic printing from Shapeways; cost of filament for your own printer (or a borrowed machine) is negligible
- Difficulty: Easy to moderate
- Safety Issues: 3D printers and laser cutters can be hazardous—if you have access to one but have never used it before, take a class or ask someone more experienced to help you

get started. See "3D Printing Linkbox" on page 72 for other titles that explain how to use these machines in more detail.

What You Need to Know

- Skills You Already Have: Drawing, cutting, assembling a model
- Skills You Will Learn: 3D computer modeling

Gather Your Materials

- Pen and paper
- Computer and access to 3D CAD software or online tool
- Access to 3D printer or online 3D-printing service

Directions

Step 1: List Your Requirements

When designing your own version of a wheel-leg, its appearance will depend on several factors:

Where it needs to operate

Think about what you want your wheel-leg to do. Will it be traveling over hard ground or soft? The feet of a wheel-leg meant for rocky hillsides or rubble-strewn city use should be sturdy but small enough to avoid getting caught in holes. In deep sand or mud, on the other hand, a larger footprint will act like a snowshoe and distribute the robot's weight over a larger surface area to keep it from sinking into the soft surface.

How it will be incorporated into a robot body

If you're going to test your wheel-leg on a pre-existing robot or rolling toy, make sure it matches the dimensions of its regular wheels and can attach to the axle or gears.

How it is built

A 3D-printed wheel-leg has to fit on the printing platform and work within the limitations of the softened plastic filament it will be made out of.

Step 2: Plan Your Project

To create an original 3D model, you may want to sketch out a rough idea of what you want first—by hand on the back of a napkin is fine at this stage. Written notes and labels will help clarify your idea. Look at photos of existing wheel-leg designs online for inspiration.

Next you need to choose the type of 3D-modeling software you want to try. There are several popular programs you can access for free, including Autodesk 123D, 3DTin, and SketchUp. Each works a little differently, but once you have used one, it's easier to learn others. You will use the free online program Tinkercad for this project because it is designed for children and other beginners and works with 3D shapes, eliminating the need to turn a 2D drawing into a 3D model.

How to Use CAD Software

CAD software has been around for a long time. Powerful (and expensive) versions for professionals are used by architects, engineers, and designers of every sort. It is also used by illustrators and animators to make two- and three-dimensional drawings and videos. Computer games are built using 3D modeling software. If you ever created a world in Second Life or built a castle in Minecraft, you've used a type of CAD software.

With the advent of 3D printers, laser cutters, and CNC routers, CAD software has become more popular than ever. Today you have a choice of several programs, each of which work somewhat differently. Many of them offer a basic version you can use for free, as well as a paid version with additional features. Some can even be used right online, just by signing up.

3D-modeling software lets you see exactly what your design will look like while you're drawing it. Depending on the program you use, you can start with a 2D representation and then "extrude" it to pull it into the virtual third dimension on your screen, or build directly in 3D. Typically the design sits on a grid that looks like the graph paper you may remember from high school geometry. Just like in geometry class, the program shows you where your design is located using coordinates on the horizontal x-axis and vertical y-axis. The z-axis runs parallel to both of these, poking out above and below the plane of the grid into the third dimension. The program also lets you change your point of view in relation to the grid, so you can view it from any angle.

Starting a new design from scratch is as simple as assembling basic shapes (e.g., square or circle in 2D, cube or sphere in 3D). You can drag the shapes from the sidebar and drop them on your grid background or "floor." Keep adding shapes, connecting them by making them touch or overlap. Some programs also let you upload a scan of a drawing, 2D digital design or photograph, or 3D scan of a model. You can then sculpt your shapes by stretching, squashing, or otherwise deforming them. Corners and edges can be beveled using the chamfering function, or rounded off with the fillet tool. To create an opening or indent, switch a positive shape into a negative space. You can also use a negative space as a tool to drill a hole or slice away a section of a shape.

Once you've created your 3D object, the discrete shapes used to create an arm or a torso can be joined together into a group. A group of shapes can be cloned, flipped, and otherwise manipulated as if they were one big shape. One group of shapes can also be subsumed into a bigger group, creating multiple levels. Ungrouping reverses the process, allowing you to back down through the levels until you are working with the individual shapes again.

When you're finished, your design can be downloaded in a file format such as STL, which is compatible with 3D printers and online 3D-printing services. You can print out your design at home, your local makerspace, or even some neighborhood copy shops. You can also send it to an online 3D-printing service like Shapeways or Ponoko. Or share your design with friends or the public on sites like Thingiverse.

For other types of 3D construction, some CAD programs will also slice your object according to your specifications, creating layers that can be reassembled into an approximation of the original 3D shape. The 3D software will create printable patterns, much like a dress pattern, to show you how to cut the slices out. These patterns can be laid out on paper, wood, Plexiglas, or other flat material to be cut out by hand, a CNC router, or a laser cutter. The cutout pieces can then be stacked or fitted together to make a good physical approximation of your design.

Step 3: Stop, Review, and Get Feedback

Before you start building, review what you've done to make sure it fits your criteria. If you can, ask a knowledgeable friend to look over your design for any obvious problems you may have missed.

> *Be sure to include some way to attach your wheel-leg to a vehicle or robot body for testing. For comparison, you can look at the wheel-leg I created on **Tinkercad** (http://bit.ly/robo-leg). You can also see it on **Thingiverse** (http://bit.ly/wheel-leg) and **Shapeways** (http://bit.ly/roboleg).*

Tinkercad Cheat Sheet

Tinkercad is a free online CAD program that lets you build designs from three-dimensional shapes. Beginning tutorial videos help you get started, and more videos are available on Tinkercad's YouTube channel—the Shortcuts video (*http://youtu.be/erEUtG8SejE*) is particularly helpful. You can also find information about design features on the Tinkercad blog. Although Tinkercad is relatively easy to use, skimpy Help and Search functions are weak points.

Here are some basics:

- Create a free account so you can save your work and share it with others.

- You can create "projects" in which to store several "designs." When you start a new design, Tinkercad will assign it a random nonsense name, which you can change by going into Properties from the Design menu. You can also save or duplicate your work from that same pull-down menu. That's useful if you want to create a backup that makes it easier to go back to a certain point and start again.

- When you create a new Tinkercad design, you will see an empty Workplane. This is the virtual surface on which you will "build" your object. Surrounding it are menus and icons you will be using as you build. In addition, various pop-up windows will appear as needed.

- To view the Workplane from a better angle, use the gray navigation icons on the left side of the screen to turn it or tilt it up or down. To zoom in on a particular piece, select it and then click on the little drawing of a box above the + and − signs. To go back to the center of the grid, click the little house icon in the center of the circle with the navigation arrows. You can also change your view by right-clicking your mouse to grab and rotate the grid. Scrolling your mouse wheel will zoom in and out on the grid.

- The sidebar to the right of the Workplane has tabs that open when you click on them. "Geometric" has a selection of premade shapes that you can click and drag over to the Workplane. "Shape Generators" lets you create your own shapes (see the directions for this project). "Helpers" includes a ruler, which you will be using in this project. If you want to delete an object once it's on the Workplane, just click on it and hit the Delete or Backspace key on your keyboard.

- "Snapping" an object to the grid lines it up to the horizontal and vertical markings on the Workplane. By default, those markings are 1.0 millimeter apart. To make more exact adjustments, you can change the Snap Grid setting (in the bottom righthand corner of the grid) all the way down to 0.1 millimeters. You can also simply turn it off.

- If you will be 3D printing your object yourself, you may be able to use one of the

presets to adjust the grid on the screen to the size of your printer's print bed, using the Edit Grid button on the bottom right.

Step 4: Design and Build Your Prototype

For this project you will make a wheel-leg from three rectangular solid legs connected in the shape of a "Y." Each leg ends in a rounded foot, and they are connected in the center by a ring-shaped hub. The final size of the entire wheel-leg is only around 60 millimeters (a little less than three inches) across, so you can print more than one at a time on most print beds. Just a note—the default unit of measurement in most 3D CAD programs is millimeters.

1. To build your wheel-leg, start with a box shape. In the menu column on the right, find the Geometric tab. Click on it to reveal a palette of shapes. Find the shape labeled "Box" and drag it over to the grid. Notice that each corner of the box is marked with a tiny white rectangle, and each side of the box has a black dot in the center of its bottom edge. When you mouse over these spots, the length of that side is shown. By default, the box shape is a cube, and each side is 20 mm.

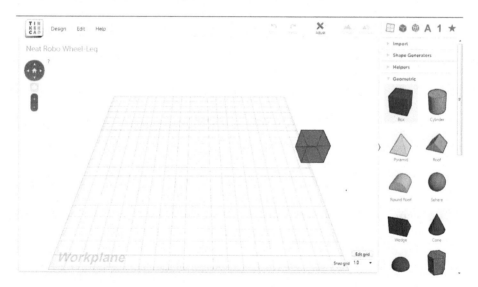

2. The basis of the leg is a rectangle 30 mm long by 5 mm wide by 8 mm high. To modify the size of the box, click on a black dot on the front to grab it. Drag it forward until it is 30 mm long.

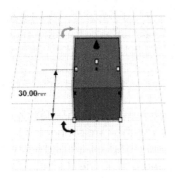

3. To adjust the width, grab the black dot in the middle of the long side you just created. Move it inwards to squeeze the box to a width of 5 mm. To adjust the height, grab the white dot at the top of the box and drag it down until the rectangle is 8 mm high. Once you've got the leg/rectangle to the right size, go to the Inspector box in the upper right of the grid and click to lock the transformation. That will keep you from accidentally changing the dimensions as you move things around.

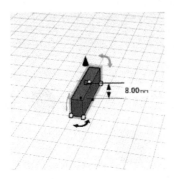

4. Now for the foot. When finished, the foot will be rounded on the bottom with a pointy toe and flat top. Its dimensions will be 20 mm long, 5 mm wide, and 8 mm high. To create this irregular shape, you'll use a shape generator, a program that lets you set the parameters of your object. From the Shape Generators menu in the sidebar (which is above the Geometric menu in the sidebar), drag an Extrusion (which looks like a gray cylinder) over to the bottom of the leg, on the left side. Viewed from above, the round gray shape should overlap the rectangle so that together they look like a lowercase "d." Before you start to modify the cylinder's round shape, first shorten the *height* to 8 mm, using the white dot on the top.

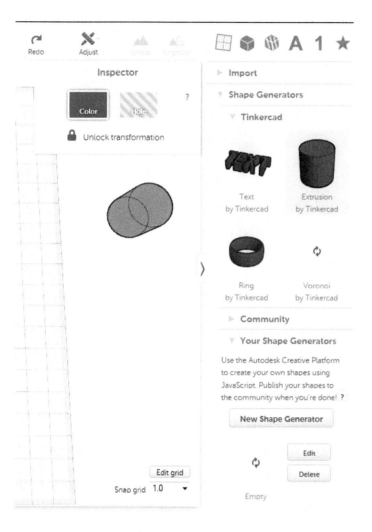

5. Next, to change the shape from a cylinder into a foot, you'll work with the smaller two-dimensional Profile grid that popped up when you clicked on the Extrusion. On it you should see an outline of a circle, which represents the cylinder as seen from above. The four dotted lines tangent to it on the top, bottom, left, and right are the Bezier handles. You use them to stretch, squash, or twist the outline. Once you've got the shape as close to finished as you can get in Profile, you can tweak it still more on the Workplane (the main grid). But before you make any changes, mouse over the little gear in the top right corner of the small Profile grid. A box labeled "Snap" will appear; check that box to make the shape line up with the grid.

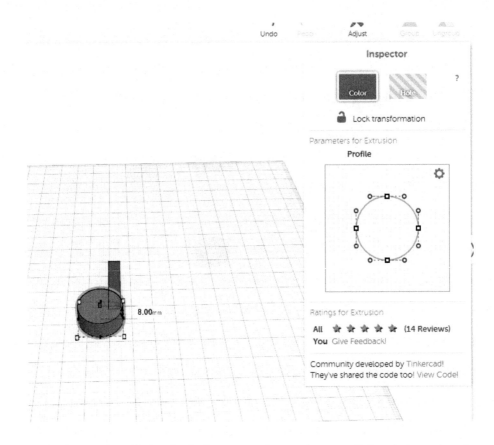

6. Grab the top handle by the little square dot at the top of the circle and pull it straight down so that it lines up with the square boxes on the side of the circle. If you wait a second, you'll see the 3D Extrusion on the big grid take on the same kidney shape as the outline on the little Profile grid.

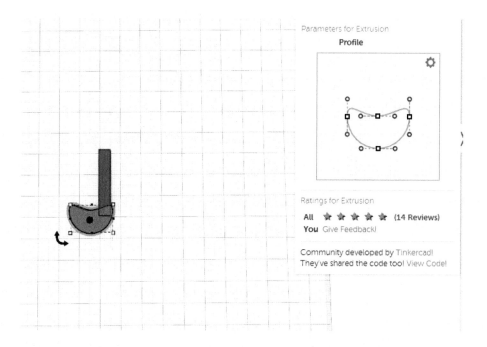

7. Grab one of the round dots on the ends of the left handle. Shorten the handle by bringing the dot towards the middle. The other dot will mirror your movements until both round dots are overlapping the middle square dot. You have now created the pointy toe.

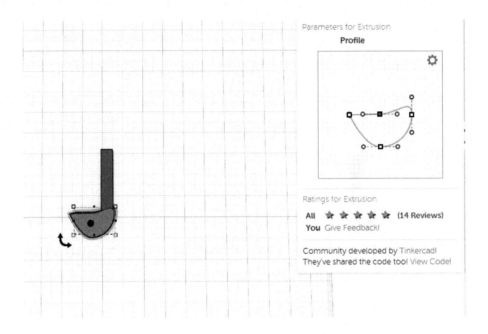

8. Grab the bottom handle by the square dot in its center. Move the handle straight up until it is level with the lower round dot on the right handle.

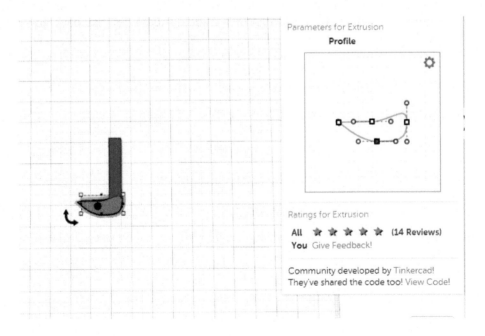

9. Center the right handle between the top and bottom handles by moving it down (grabbing it by the square dot). Shorten the right handle to fit between the top and bottom handles. Next, tilt the right handle until the top round dot covers the rightmost round dot on the top handle.

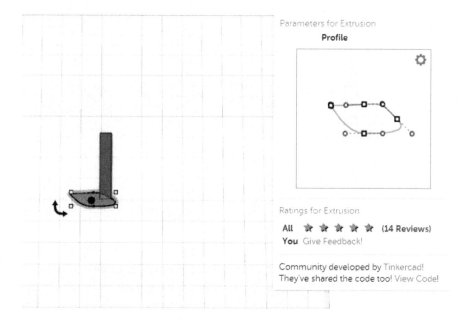

10. Lengthen the bottom handle by dragging the right round dot until it is directly under the square dot on the right handle. Then grab the bottom handle by the square dot and move it to the right until it connects with the right handle. You may want to uncheck the Snap function to get it straight.

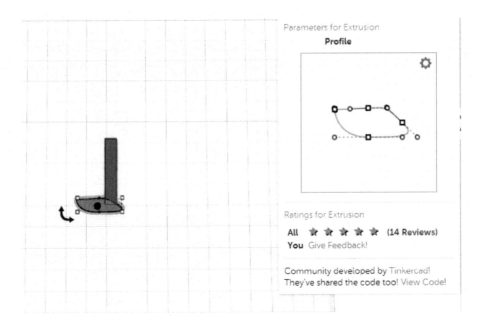

11. Going back to the Workplane, the foot should be roughly shaped like a parallelogram. Move the foot if needed so it overlaps the rectangular leg at a right angle. You want a smooth "sole" on the bottom of your foot, so take a look from that viewpoint to make sure the leg isn't poking through the "heel." If you want, you can sharpen the corners of the foot by lengthening the handles on the Profile grid so they overlap slightly. Or you can point the toe or the foot a little up or down by rotating it on the Workplane using the curved double-headed arrow that appears when you click on the foot.

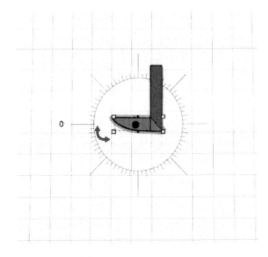

12. When you are satisfied with the design, open up the Helpers menu (right above the Geometric menu) in the sidebar, grab the Ruler and drag it over to the foot. The measurements of the foot should appear. You can change the measurements as desired by simply typing over the numbers. Make the foot 5 mm wide and 20 mm long. If needed, slide the foot around one last time to get it in the right position. Dismiss the ruler by clicking outside the foot, and then on the gray *X* circle.

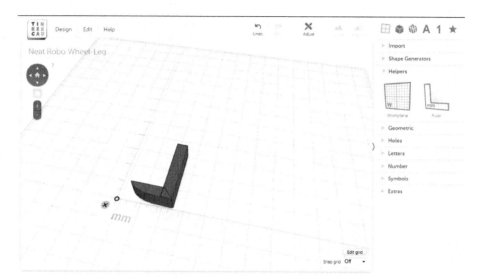

13. Finally, select the entire leg by clicking on the grid and dragging the light blue outline around until the entire leg is inside. Click Group on the top right of the page to save the leg-foot shape as one piece and avoid changing it accidentally. The leg will switch to a single color. (You can also change the color yourself at any time with the Inspector pop-up box.) If you need to work with the pieces individually later, you can always ungroup them.

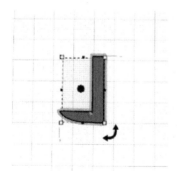

14. Now you can make the other legs. Click on the leg, then click on Edit at the top right of the page, then Duplicate. Click on the leg to slide the duplicate away from the original and up to the left. Then rotate the second leg 120 degrees, using the protractor that appears when you mouse over the curved double arrow. Move the rotated leg so its corner touches the top corner of the first leg like a hinge. If necessary, turn off the Snap grid function to give you more control. Duplicate and rotate the second leg the same way to make the third leg. Move it so that it touches the other two legs. The tops of the three legs together should form an equilateral triangle at the center. Zoom in as much as possible to check that the three edges are touching, without gaps or overlapping. Then group the three legs together.

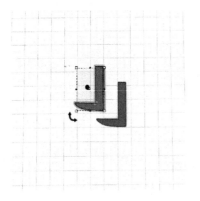

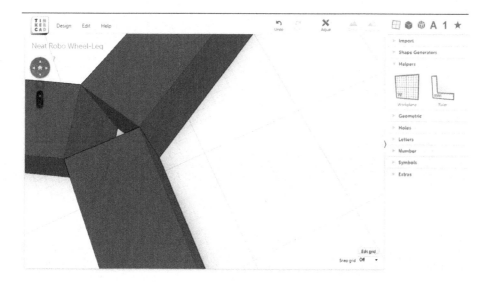

15. The last step is to make the hub of the wheel-leg. Go to the Geometric menu and drag a cylinder onto the grid to work on it, a little way from the wheel-leg. Resize the cylinder to 8 mm high, 12 mm long, and 12 mm wide. Then go to the Holes tab in the right sidebar and drag a Cylinder Hole onto the grid. Set it to 10 mm high to be certain that it goes completely through the hub. Set the diameter of the hole to 3 mm. That should fit on an axle made from a bamboo skewer. (To adjust the size of the hole, see the section on troubleshooting.) When finished, drag the hole into the solid hub. Poke the hole down below the grid a little bit so the ends stick from the top and the bottom of the hub. To center it, select the cylinder and hole, go to Adjust on the top menu bar, and click on Align. On the grid, guidelines will appear. Click on the guidelines that cut through the middle of the cylinder horizontally and vertically. Then dismiss the Align tool, select the pieces, and group them together.

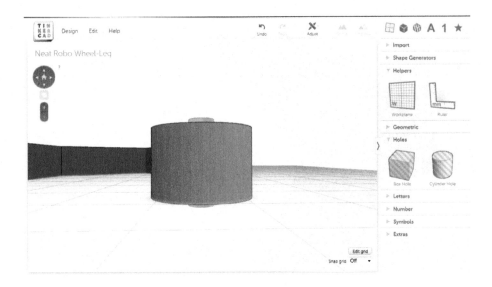

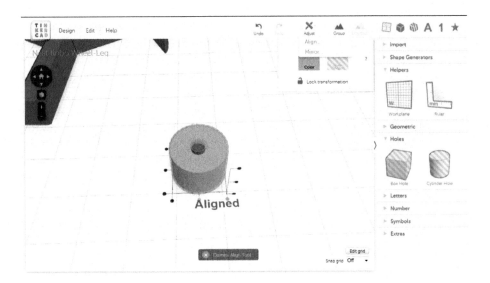

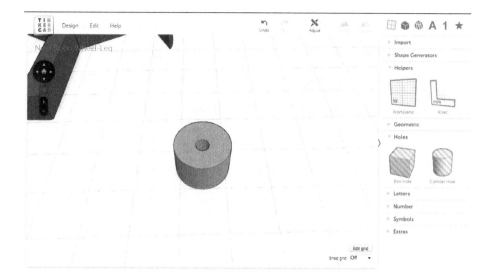

16. Finally, drag the hub into the middle of the wheel-leg. The Align tool will not work here, so center the hole as well as you can by eye. Group all the pieces together. Remember, you can always go backwards and ungroup them, layer by layer, if need be.

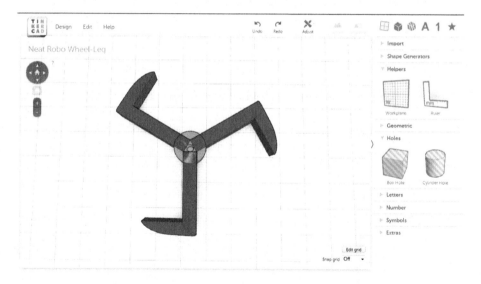

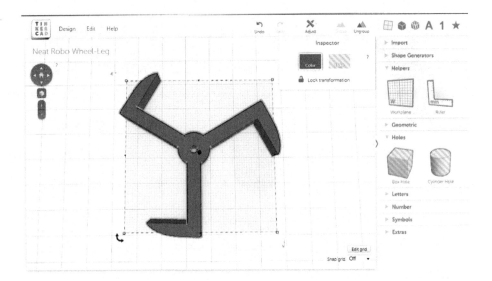

17. To download your wheel-leg for 3D printing, choose that option from the Design menu at the top right of the page. You will be asked what kind of file to save it as (this depends on the 3D printing machine you use). The Design menu also contains links to let you upload your file to Thingiverse or order a 3D print of your creation from online services such as Shapeways, i.materialize, Ponoko, or Sculpteo. You can also download the file for laser-cutting.

Step 5: Test Your Design

When your printed wheel-legs are ready, attach them to a bare-bones platform to test them out. You can make axles out of cocktail toothpicks, bamboo skewers, or wooden dowels. Poke them through one of the "tunnels" inside a piece of corrugated plastic or cardboard, attach the wheel-legs, and make them move by placing them on a tilted surface, pulling them with a string, or adding a motor. You can also try out your wheel-legs by substituting them in for wheels on a toy car or robot platform you have available.

Step 6: Troubleshoot and Refine

Most hobbyist-quality 3D printers are not that exact when it comes to tiny details like the hole in the hub, so getting the wheel-leg to fit on whatever you use for an axle can take a little trial and error. For a better grip, you can try making the circular hole more "D" shaped by overlapping a small box on one side. You can also turn any shape into a hole using the Inspector pop-up box. A star-shaped hole with points to grab a wooden shaft may allow for more variation in the thickness of the axle, for instance. Or, a less high-tech solution is to widen the hole once the wheel-leg is printed by carefully hammering in a nail of the right diameter.

If your wheel-legs can spin freely around the axle but don't roll smoothly along the ground, you may need to go back and look at your design. Do the feet lie along the line of an imaginary rim? If not, your "wheel" may not be round enough. You can also try making the feet shorter or turning up the toes a bit more. Use the results of your tests to modify your drawing and make a revised prototype.

If the feet slip while trying to move on slick surfaces, you can give them some traction with a peel-and-stick craft foam "shoe," or put treads on the bottom of your wheel-leg feet with lines of hot glue. You can also cut out the ridged fingertips from a rubber dishwashing glove and use it for the sole of the shoe.

Step 7: Adaptations and Extensions

Once you've got the hang of using the 3D CAD software, and have tested out the basic design, try creating a wheel-leg prototype of your own. Questions you should consider include:

- How many legs per wheel work best?
- Are legs with bent knees better than straight legs?
- How much curvature should the feet have?

If you want to try making the wheel-leg by hand, using cardboard or other material rather than 3D printing, use a CAD program that lets you design in 2D as well as 3D, or which lets you convert your 3D model into slices. Programs to try include AutoDesk 123D and SketchUp.

To extend the project, you can go on and design the rest of the robot using 3D modeling, to find the best body and attachment options for your wheel-leg design.

3D Printing Linkbox

Bob the BiPed (*http://bit.ly/bob-biped*)

Jimmy, the 21st Century Robot (*http://robots21.com/*)

Robohand (*http://robohand.net/*)

e-NABLE (*http://enablingthefuture.org/*)

MakerBot (*http://www.makerbot.com/*)

RepRap (*http://reprap.org/wiki/RepRap*)

AutoDesk 123D (*http://www.123dapp.com*)

SketchUp (*http://www.sketchup.com*)

Tinkercad (*http://www.tinkercad.com*)

3DTin (*http://www.3dtin.com/*)

Thingiverse (*https://www.thingiverse.com*)

Shapeways (*http://www.shapeways.com*)

Ponoko (*https://www.ponoko.com*)

Sculpteo (*http://www.sculpteo.com/en*)

i.materialise (*http://i.materialise.com/*)

Wheel-leg on Tinkercad (*https://tinkercad.com/things/88TMrSsQTR2*)

Wheel-leg on Shapeways (*http://bit.ly/roboleg*)

Make: 3D Printing: The Essential Guide to 3D Printers (*http://bit.ly/make-3d-printing*) by Anna Kaziunas France (Maker Media, 2013)

3D Modeling and Printing with Tinkercad: Create and Print Your Own 3D Models by James Floyd Kelly (Que, 2014)

Unevolved Robots | 3

There are minimal elegant solutions to building capable devices so long as they spend their energies on themselves and not their "masters."

— Mark W. Tilden

How simple can a robot get and still be considered a robot? That's a question many robot makers have taken as a challenge. Most people would agree that a doll or a puppet that only moves when pulled or pushed by a person is not a robot. But how about a remote control model rover, which adds a layer of electronics between the human and the object? Is something a robot if its sensor is a mechanical on-off switch rather than a logic gate in a computer program? Or if its seemingly smart response to the environment around it is really a random series of movements? At the beginning of the computer age, it seemed that robot brains would keep growing in processing power. Instead, the challenge today is to see how many different situations a robot can handle with the least number of possible responses to choose from.

The earliest precursors of today's robots didn't have computer brains or electronic sensors. In fact, they didn't even run on electricity. Automata—mechanical robots powered by wind-up mechanisms, springs, and weights—go back as far back as ancient China. Leonardo da Vinci designed a mechanical soldier in 1464. Swiss watchmaker Pierre Jaquet-Droz built a writing automaton in the shape of a boy in the late 1700s that still works today. Using a system of interchangeable teeth on a wheel that controls gears inside, it can be programmed to write up to 40 characters. As it writes, its eyes follow the movement of the quill pen across the page, and its hand stops to dip its quill pen in an inkwell. The Franklin Institute in Philadelphia houses its own drawing boy with the largest "memory" of any automaton known. It is programmed to create four drawings and three poems, two in French and one in English. Built by another Swiss clockmaker, Henri Maillardet, around the same time as the Swiss machine, it was the inspiration for the automaton seen in the movie *Hugo.*

Today, animated figures or entire scenes that move with the turn of a crank are found on wind-powered lawn ornaments and as DIY paper models. You can buy spring-operated dolls that walk, hop, and even flip in the air at any novelty shop. Slightly more upscale, Brazilian artist

Chico Bicalho has a line of metal and wire mini-insectoid robots (Figure 3-1) activated by springs that you wind up with a key or by pulling a string. They move due to the vibration of a spinning off-center weight.

Figure 3-1. *Chico Bicalho's vibrating robots are powered by a wind-up spring.*

Then there's Dutch artist Theo Jansen's Strandbeests, complicated pipe structures that use an ingenious series of joints (Figure 3-2). The original Strandbeests, built on a North Sea beach, were as large as elephants (hence the name, from the Dutch for "beach animal"). Made of PVC plumbing, they were propelled by wind, which they captured with sails and stored in a series of "stomachs" made of recycled plastic bottles. The latest generation of Strandbeests are tabletop-sized, feature windmill blades that can be powered with an electric fan, and can be reproduced—complete with fully assembled articulated joints— on a 3D printer.

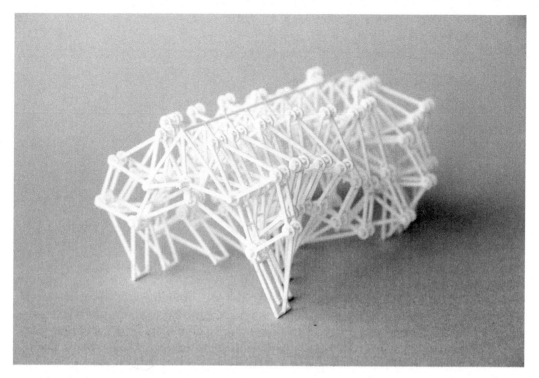

Figure 3-2. *A 3D-printed working model of Theo Jansen's wind-powered Strandbeests that can be purchased on Shapeways.com. Credit: Tim van Bentum.*

Modular robots are basically a set of parts that can assemble themselves in different ways to handle a variety of tasks. MIT researchers have developed box-shaped modular robots that move without any external actuators. Called M-Blocks, the cubes—about the size of a child's alphabet blocks—fling themselves about with the help of a spinning weight inside. When the flywheel stops suddenly, the block keeps going, leaping into the air. Magnets along the edges help it grab onto nearby M-Blocks, allowing the robots to arrange themselves into the desired shape piece by piece. A toy called Cubelets is another form of modular robot (Figure 3-3). In this case, every Cubelet (similar in size to M-Blocks) performs a Sense, Think, or Action function. You "program" the Cubelet robot through your choice of modules. For example, if you combine a Brightness sense module with a Maximum think module and a Drive action module, you get a robot that measures the amount of light around, analyzes the data to see if it is over a certain threshold, and if it is, begins to move in that direction—in effect, a light-seeking robot.

Figure 3-3. *Cubelets are modular robotic blocks used as building toys. Credit: Modular Robotics.*

Swarming robots are individuals that work in sync with each other. Inspired by social insects like termites, bees, and ants, swarming robots rely on group-think to accomplish their task. Harvard's TERMES robots can build complex structures out of foam bricks with only four sensors, three actuators, and minimal brainpower to guide them. No bigger than a dinner plate and equipped with wheel-legs to help them climb, the TERMES robots perform a simple yes-no test on their immediate surroundings to decide whether the spot they're at needs another brick or not

This programming lets them operate independently without having to follow a master plan and still create designs like towers, castles, and pyramids.

This chapter will show you how to build two electronic robots from the lower branches of the family tree: a Gliding Vibrobot and a Suped-Up Solar Wobblebot.

Unevolved Robot Linkbox

Franklin Institute Maillardet's Automaton (*https://www.fi.edu/history-automaton*)

Chico Bicalho's wind-up vibrating robots (*http://www.kikkerland.com/designers/chico-bicalho/*)

Theo Jansen's Strandbeests (*http://www.strandbeest.com/*)

M-Blocks (*http://ppm.csail.mit.edu/node/63*)

Cubelets (*http://bit.ly/cubelet-kit*)

TERMES (*http://bit.ly/termes-proj*)

Project: Make a Swarm of Gliding Vibrobots

What Is a Vibrobot?

A vibrobot, or vibrating robot, is a primitive bot that moves due to the shaking of a rotating eccentric (off-center) weight, usually a vibrating motor.

Figure 3-4. *Three Vibrobots swarming.*

What It Does

Vibrating robots behave in lifelike ways without any electronic control through a combination of random movements and mechanical programming due to their shape and distribution of weight.

Where It Came From

Probably the most famous example of a vibrating robot is the Bristlebot—the Throwie of the robot world. First appearing on the Evil Mad Scientist Laboratories website in 2007, the Bristlebot consists of the head snipped off a toothbrush, a tiny vibrating motor like those used in pagers and cell phones, and a disc battery. The bristles direct the bot in a roughly forward direction, with just enough unpredictability to give it character. Hexbug's popular Nano robots are basically Bristlebots with added decorations and sometimes sensors. Another well-known vibrating robot design is usually known as the ArtBot. This has a larger DC motor to which a weight can be attached. The legs consist of felt-tip markers, so it draws as it moves. (You can find directions for a customizable Art-Making Vibrobot in my book *Robotics: Discover the Science and Technology of the Future*.)

However, not all vibrobots are "dumb." Kilobots are tiny vibrating robots developed at Harvard. They can be programmed to move in preset patterns or to head towards or away from each other. About the size of quarter, each Kilobot consists of a circuit board, battery, two vibrating motors, and infrared sensors set upon three toothpick-like legs. Infrared signals (like the signals your remote sends to your TV when you push the buttons) tell the Kilobots what to do, and can program dozens or even hundreds of Kilobots simultaneously. The Kilobots can also send infrared signals to each other. Each Kilobot can be built from less than $50 worth of parts, and someday soon it may be possible to manufacture Kilobots in bulk for around $15 a piece, so assembling a Kilobot army won't break the bank. In the meantime, you can buy a 10-pack of pre-assembled Kilobots for research or education purposes for around $1,000.

A swarming robot called I-Swarm was announced by a consortium of European scientists in 2009. Tinier than a pencil eraser, I-Swarm moved about on three tiny legs via sound vibrations and was steered by changing the frequency of the vibration, just like the Union College tensegrity robot in Chapter 2. They even hummed as they turned and glided around. Although the I-Swarm project never got beyond the prototype stage, even smaller microrobots could one day be used to deliver medicine or perform procedures inside the body.

How It Works

In 2013, Italian researcher Luca Giomi used custom-designed Bristlebots to show that collective behaviors like swarming can arise simply by letting self-propelled devices interact in a confined space. Working with a team at Harvard, Giomi developed two types of BBots (as they were called)—walkers and spinners—by changing the angle of the bot's bristly feet. The study found that when a crowd of each type of BBot reached its particular critical mass, the bots would start to move in sync, like a school of fish or flock of birds. The finding suggests that to some extent swarming behavior may be a result of the physical dimensions of a group as well as hardwired or voluntary decisions on the part of the individual swarm member. The study called this behavior the "mechanical intelligence" of swarms.

Making the Project

The Gliding Vibrobot is a Bristlebot reduced to its bare essence: just a vibrating motor, a disc battery, and a smooth round container to hold them. Although the Gliding Vibrobot doesn't have feet that let you program it to move in a particular direction, its random wanderings still seem to result in that natural ability to swarm found in the research at Harvard. Build at least three and watch how they tend to form a pack if left long enough around a limited test area.

 Don't forget to document your work!

Project Parameters

- Time Needed: 1 hour
- Cost: $10 or less

- Difficulty: Easy
- Safety Issues: None

What You Need to Know

- Skills You Already Have: Basic craft skills
- Skills You Will Learn: Observation, trial and error

Gather Your Materials

- Three or more small vibrating motors (available from most electronics dealers, or recycle one from an old cell phone, pager, or Oral B Pulsar disposable electric toothbrush)
- Three or more small coin cell batteries (1.5 to 3 volts)
- Foam tape
- Three or more 2-inch (5 cm) acorn-shaped plastic gumball machine capsules (the kind used to dispense toys)
- Smooth test surface, such as a large piece of poster board and/or cardboard box top for a confined test area

Directions

Step 1: List Your Requirements

The goal of this project is to assemble a quick-and-easy Gliding Vibrobot that travels in large loops. As the number of little bots grows, so will the patterns you are likely to see. The more the merrier!

Step 2: Plan Your Project

Putting the pieces of this project together is easy. The challenge comes in positioning them just right, to achieve the kind of movement you want to see. While the plastic gumball machine capsule looks neat without adornment, you can add personality—and change the behavior of your bots—by adding craft decorations (see Step 7). Keep in mind that this project has no on/off switch. You should probably have a few extra batteries on hand, since you will burn through them as you test (and play with) your Vibrobots.

Step 3: Stop, Review, and Get Feedback

Check out the original Bristlebot how-to on the Evil Mad Scientist website for inspiration and building tips.

Step 4: Build Your Prototype

1. If you are repurposing an old vibrating motor, make sure each of its wires have a little metal exposed at the end. If not, strip off about 1/4 inch (3 mm) of insulation with a wire stripper (or carefully with a scissor or wire cutter).

2. Cut a piece of foam tape about 2 inches (5 cm) long. Stick the motor onto the tape, with the weight on the shaft hanging off the end so that it can spin freely.

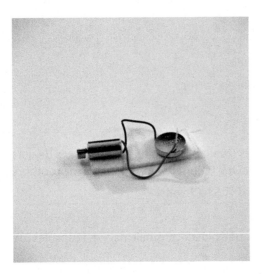

3. Press one wire down along the tape. If the wire is more than an inch (2.5 cm) long or so, curve it a bit so that the metal end is closer to the bottom of the motor. Then press the disk battery on top of the metal end of the wire to hold it in place. When you're ready to turn your Vibrobot on, secure the other wire to the top of the battery with more tape.

4. Take off the clear plastic lid on the gumball capsule and set it aside. Stick a square of foam tape inside the bottom (colored flat part) of the capsule, a little off-center.

Attach the foam tape holding the motor and battery onto this piece of tape. You will have to fiddle with it a bit to get the Vibrobot to move forward and not spin in place. One configuration that works puts the battery up against the side (for ballast) and the weight on the motor shaft almost touching the side as well.

5. Make at least two more Gliding Vibrobots the same way.

6. Start each Vibrobot moving by attaching the top wire to the top of the battery with foam tape. Snap on the clear capsule lid and you're ready to try 'em out.

Step 5: Test Your Design

1. Give your Gliding Vibrobots a clear space to move around in (Figure 3-4). A big piece of poster board makes a nice smooth surface. You can make a test area with walls by putting some inside the lid of a copy paper box.

2. Turn them on, set them down, and watch how they move.

3. Adjust the position of the motors and the batteries as needed to get the kind of motion you'd like.

Step 6: Troubleshoot and Refine

Make sure the vibrating motor in your Vibrobot is firmly attached to the inside of the bot at all times. If your bot just sits and spins, try moving the parts around until it moves forward-ish.

Step 7: Adaptations and Extensions

This ultra-simple bot is just crying out for ornamentation. Stick on some googly eyes, ballpoint pen springs for antennae, or add minicraft stick "skis" to the bottom. You can even create flying saucer landing gear by sticking some long straight pins with plastic ball ends through the bottom of the plastic gumball capsule and hot gluing in place. For advanced makers, try adding some popular Bristlebot upgrades like two flat vibrating motors and a light-sensing circuit that tells the bot to turn towards or away from the light.

Gliding Vibrobot Linkbox

Evil Mad Scientist Bristlebot tutorial (*http://bit.ly/emsl-bristlebot*)

Bristlebot swarm research (*http://bit.ly/bristlebot-report*)

Kilobots (*http://bit.ly/kilobot-project*) (order at K-team (*http://bit.ly/get-kilobot*))

ScienceBuddies Steerable Bristlebot (*http://bit.ly/light-bb*)

Hexbug (*http://www.hexbug.com*)

Project: Make a Souped-Up Solar BEAM Wobblebot

What Is a BEAM Robot?

A BEAM robot is a primitive solar-powered bot that behaves in lifelike ways.

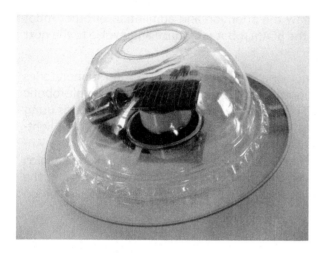

Figure 3-5. *A finished wobblebot*

What It Does

A BEAM robot is controlled by a primitive "nervous system" that uses components like switches and transistors instead of a computer "brain."

Where It Came From

For hobbyists, the bare-bones robot craze really came into its own with the advent of BEAM Robotics. BEAM—which stands for Biology, Electronics, Aesthetics, and Mechanics—was formulated in 1991 by robotics physicist Mark W. Tilden. Tilden, who worked on both NASA rovers and the Wowwee Robosapien line of toy robots, broadened the definition of "robot" to include any machine that could interact with its environment, even if its behavior relied more on reflex than reasoning. In his book *JunkBots, BugBots & Bots on Wheels* (McGraw-Hill Osborne Media), Tilden and coauthor David Hrynkiw describe the concept of *appropriate technology*, the idea of using the most low-level components possible. Among other advantages, building things with primitive technology lets you use parts that are cheaper and easier to find. It's also easier to diagnose problems, because there are fewer parts and they're less complicated to test and fix.

The *Biology* in BEAM refers to the idea that nature provides many elegant models that can be used to solve robotic design problems. But it also suggests that robots, like living things, exist along a spectrum that ranges from primitive to complex. Like any primitive creature, the BEAM robots' primary objective is to survive. Since BEAM robots are typically solar-powered, that means they can find their own food wherever there is light. For this reason, many BEAM robots have *Electronic* sensors to detect light and direct the robot towards it.

The *Aesthetics* and *Mechanical* aspects of BEAM come into play in the design. A well-designed BEAM robot is neat, tidy, and attractive. Everything fits together, so pieces are not apt to come flying off. Ideally, its design is also efficient. Just like programmable materials and smart bodies, the architecture of a BEAM robot should help it move the way its supposed to, and hinder it from doing things it shouldn't. Good aesthetics also means it's nicer to look at—which can help it draw the attention and admiration of other robot builders, who may be inspired to help the BEAM robot reproduce and evolve to the next level.

BEAM robots are mainly a hobbyist phenomena. They're not generally found in laboratories or industry, although they heavily influenced the present-day trend towards simple robotic design. You're most likely to see BEAM robots in the form of scratch-built designs using inexpensive or reclaimed components, or as kits. One well-known model is the light-chasing MouseBot, so-called because of its origins as an upcycled computer mouse. Another favorite is the elegant bump-and-turn BeetleBot designed by Jérôme Demers, which has two antennas connected to switches that act as touch sensors. BEAM was also the inspiration for the Hexbug line of miniature toy robots, many of which employ common BEAM elements.

How It Works

One classic type of BEAM power system is the Miller Solar Engine, designed by Andrew Miller around 1999. It consists of a storage capacitor—a component that works something like a rechargeable battery, except that it releases all its electricity in one burst—which is charged by the solar panels, and other components that determine when the electrical charge in the storage capacitor is released. Like the safety valve on an overflowing tub, when the charge in the storage capacitor reaches a certain level, a smaller capacitor will let electricity flow to a voltage trigger, which activates a transistor that sends power to the motor. When the storage capacitor is drained below the smaller capacitor's threshold, it goes back to storage mode until it overflows again. As a result, BEAM bots run in short bursts, that can be lengthened or shortened according to what kind of capacitor is used.

Making The Project

This project combines a Miller Solar Engine with the design of a Solar Wobblebot, a project from my book *Robotics: Discover the Science and Technology of the Future*. The original Solar Wobblebot (Figure 3-6) ran directly off the solar panel from a dollar store garden light—but only when there was enough light to power it. The Miller Solar Engine can also work in lower light levels by storing energy in a rechargeable capacitor and releasing it all at once. The stronger the light, the more frequent the bursts.

Figure 3-6. *The original design for the Solar Wobblebot*

 Don't forget to document your work!

Project Parameters

- Time Needed: 3–4 hours
- Cost: $10–$30
- Difficulty: Moderate
- Safety Issues: Soldering tools and components touching them become extremely hot and should be handled carefully. Components must be connected with the correct polarity (positive to positive, negative to negative), or they may heat up and burst.

Gather Your Materials

Miller Solar Engine Components

You can order all the electronic parts you need for this project from *Solarbotics.com* or other retailers. There are two configurations to choose from—fast, short bursts or slower, long-running releases (Figure 3-7). I like both, but you may get the most immediate gratification from the fast/short burst variety. Solarbotics offers bundles that have all the components you need, including a solar panel with a circuit board on the back, which makes soldering a lot easier. The solar panels come in two sizes, 4.5 volts and 6.7 volts; either will work. The bundles available are the following:

Figure 3-7. *Components for long-running option. Top, L-R: PN2222 transistor, 1381 voltage trigger, disc capacitor, barrel capacitor. Bottom: diode.*

- SCC3733a-MSE Fast Bundle 1 (short bursts, 6.7V solar cell)
- SCC2433b-MSE Fast Bundle 1 (short bursts, 4.5V solar cell)
- SCC3733a-MSE Powerful Bundle 2 (long-running, 6.7V solar cell)
- SCC2433b-MSE Powerful Bundle 2 (long-running, 4.5V solar cell)

    Note that the motor that comes with the bundles above is narrow for Solar Wob-blebot purposes, so you may need to jury-rig a holder from a piece of cardboard. You can also substitute one of the motors suggested further down, which have a wider body.

    If you feel like a challenge, you can also "freeform" your components together without a board and use your own repurposed solar panels from cheap solar garden lights. The components listed below are the same as in the bundles above, but you can order them separately and just get what you need. That's a good choice if you are making several bots and prefer to use your own solar cells and motors.

- PN2222 NPN transistor (Solarbotics #TR2222)

Transistors labeled PN2222 and 2N2222 are interchangeable; a third type, P2N2222, has the leads in reverse order, just to confuse us.

- 1381 voltage trigger (Solarbotics #1381)
- 1N914 silicon diode (Solarbotics # D1)

Long-running option
- 6.8µF Tantalum (disc) capacitor (Solarbotics #CP6.8uF)
- 0.35F 2.5V 5X11.5 mm (storage/barrel) capacitor (Solarbotics #CP0.35F)

Short-burst option
- 1.0µF monolithic capacitor (Solarbotics #CP1.0uF)
- 4700µF electrolytic capacitor (Solarbotics #CP4700uF)

Solar panels—you only need ONE of the following choices!
- Solar cell/Miller Solar Engine circuit board (Solarbotics #SCC2433B-MSE)
- Two dollar store solar garden lights
- Solar cell rated at 4 volts or above

Motor—you only need ONE of the following choices!
Note: If your motor does not already have wires attached, you will have to solder some on.
- Low-speed DC motor (Home Training Tools #EL-MOTOR2)
- Mabuchi RF500TB DC motor (Solarbotics #RM4)
- Low-voltage (1.5V or less), low-inertia DC motor (may be sold as "solar motor")
- Recycled low-voltage DC motor from old cassette tape, VCR, CD, or DVD player

Tools
- Hot glue gun
- Wire cutters and/or wire strippers
- Glue dots
- Solder and soldering tools (see "Soldering Cheat Sheet" on page 88)
- Needle-nose pliers
- Heat shrink tubing and heat gun (a crafter's heat tool is fine) or electrical tape

Solar Wobblebot body
- Old CD or DVD disk
- Clear plastic drink cup dome, or other food container (like that for small hummus dips or salad)

- Pizza box spacer, fast food ketchup cup, or other platform for attaching the components to the body above the motor (if needed)
- Eraser from a standard pencil (if desired)

What You Need to Know
Skills You Already Have
Following assembly directions, reading diagrams

Skills You Will Learn
Soldering, assembling electronics, making an electrical circuit

Soldering Cheat Sheet

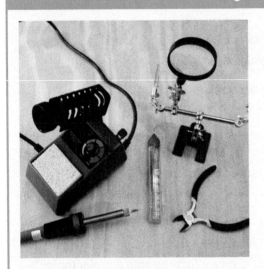

There's a lot to know about soldering, and it takes a bit of practice to get good at it. That said, a beginner can do a good enough job for the projects in this book, even if what results isn't pretty. Your goal in soldering is to make a secure bond between the parts you want to connect, without creating accidental bridges between the parts you *don't* want to connect—all while avoiding frying yourself, your furniture, or your components. Here's what you need to know to get started.

Supplies
These are the bare-bones basics. See "Soldering Linkbox" on page 91 (and related print resources) for more detailed advice about soldering tools and supplies. (I highly recommend the Make: Getting Started with Soldering Kit from Maker Shed. It comes

with everything below except safety goggles, and includes a copy of Make's *Learn to Solder* book.) There are desoldering and other tools that you will need as you become more skilled and do more advanced soldering projects. The small tube of solder included in the kit is adequate for all the projects in this book, but you may want to get an additional roll of rosin-core solder (see below).

- Soldering iron: A pencil-type soldering iron with a thin pointy tip will help you put the heat just where you want it. Inexpensive models come in 15, 25, and 30 watts. The lower the wattage, the longer it will take to heat up, but the less likely you are to fry any of your components.

- Solder: For electronic components, a good general size to use is 0.025 inch (0.6 mm). The thinner the solder, the easier it is to melt. Be sure to use rosin-core solder. Rosin is a substance that melts onto and cleans the metal as you work, helping to create a good electrical connection. Many electronics veterans still prefer lead solder, because it has a lower melting point than lead-free varieties and is easier to work with. Lead-free solder has its own risks: it releases particulate-laden fumes when heated. However, if you're working with children or will be handling the soldered parts once it's done, then lead-free solder is your best bet. See below for solder safety tips.

- Helping hand tool: This is a stand with bendable arms that have alligator clips on the ends. It can be incredibly handy (ahem!) when you need to keep your components steady while you're working with the hot soldering tool. Some come with a magnifying glass to help you see what you're doing. Some come combined with the soldering tool stand (see below), but if you're tempted to buy one of those, check the reviews to make sure it doesn't tip over if you try to swing the magnifying glass into a usable position. (I'm looking at you, RadioShack.)

- Soldering tool stand: If you bought a beginner set with a little bent piece of metal to lean your hot iron against, consider getting a separate stand. Look for one that will hold even a narrow soldering tool up off the work surface. A place for a sponge is nice too (see next tip).

- Damp sponge: The metal tip of the soldering iron reacts with the surrounding air whenever it gets hot, causing a layer of oxidized metal to form. So you'll need to clean it while you're soldering by wiping the hot tip across the sponge, often while you're working. If your stand doesn't have a sponge holder, make one out of an Altoids tin or metal jar lid. Cut a plain flat sponge to fit, then dampen it and place it in the holder. Keep it with your soldering supplies—you *do not* want to accidentally use the same sponge on your dishes.

- Desoldering braid: This is a loosely woven strip of ribbon made of extremely thin copper wire. To remove excess solder from a joint, put a piece of braid over it and press it down with the hot soldering iron. The braid will wick away some of the melted solder from the joint. Lift it off the joint while it is still hot. Be careful not to touch the hot braid— hold it by the spool or use your pliers (below).

- Safety goggles: Safety goggles will prevent flying bits of wire or solder from landing in your eye.

- Needle-nose pliers: Here's some truly quick-and-dirty advice for beginners who are having trouble making nice solder joints. If your wires are not quite soldered together the way you'd like, you can use your pliers to gently squish them together well enough to hold. You can then try to carefully add a little more solder— but be aware that as soon as you heat up the wire again, you run the risk of loosening up the existing solder. And when it's easier to work with the components sitting right on the work surface, you can fit them into the open jaws of the pliers to hold them steady instead of clipping them onto a third hand tool.

Soldering Steps

1. Protect yourself. As mentioned above, solder contains lead and other nasty substances (even lead-free solder gives off toxic fumes), so use good ventilation. A small tabletop fan or Make's DIY Mini Fume Extractor (see link below) can help, especially when you can't open a window. If using lead solder, be aware that it's most dangerous when ingested, so be sure to pick up and dispose of any leftover bits and wash your hands when you're finished or before touching food. And take care with the hot soldering iron. Plug in the electric cord where no one will

trip over it, and turn it off whenever you step away.

2. Make your work area safe. Do your soldering away from the family living space whenever possible—in the basement or garage is ideal. If that's not an option, cover your workspace to make it easier to collect stray bits of solder and wire and be sure to work on a sturdy, heat-resistant surface. When I can't get down to my local makerspace to solder, I throw an old vinyl tablecloth over my dining room table and set up my tools on a repurposed glass cutting board or slab of plywood.

3. Make sure the soldering iron is hot enough. Most low-cost soldering irons don't have temperature controls—they're either on or off. To be sure your soldering iron is hot enough, rub it on the damp sponge. It should cause the sponge to steam, and any solder it touches should melt within a second or two.

4. Tin your soldering iron's tip. Once the soldering iron tip is clean of old solder, you can protect it by adding a new protective layer. This is called tinning the tip. Briefly touch the solder to the tip. Make sure to cover all around the tip. Wipe the tip gently across the damp sponge until it the new layer of solder is smooth and shiny. Repeat the tinning process as needed while you're working.

5. To make a good joint, heat the thing(s) you want to solder. Solder flows towards heat. So heat up the wire, metal pad, or component you're working with before adding the solder. To do this, hold the tip of the hot soldering iron to the place to be soldered for a second. Then push the solder under the tip until there is enough to cover the joint without blobbing over onto any nearby connections. Remove the solder but continue to hold the soldering iron to the joint for another second. This lets the solder flow smoothly around the joint. When soldering two wires together, the solder should form a little smooth mound around them. When soldering a wire through a hole in a circuit board, the solder should flow into the shape of a volcano, with the wire sticking out the top. After removing the soldering iron and letting the joint cool for a few seconds, try to wiggle the pieces around to make sure they are secure. If you need to add more solder, be very careful—once the soldering iron heats the existing solder, the joint may come apart again.

6. To move the solder from one spot to another, heat the new spot. If your solder overflows onto another joint, you create a *solder bridge* that can short your circuit. Move the excess back to its rightful place by applying heat to the spot where you want the excess to retreat to. Or use the desoldering braid to remove the excess altogether.

7. Be careful not to overheat components. Some components are touchy and will burn out if overheated. It's also possible to melt a circuit board. Touch the soldering iron to your components as briefly as you can to make a good joint. And if you've got a piece of shrink tubing on the wire ready to slip over the finished joint, make sure it's well back from the heat, or it may shrink before you're ready. (If that happens, you may be able to stretch it open with a pin.)

Break the Code: Magic Smoke

Overheat your components while soldering and you may see what veterans call "magic smoke" drift away. The caustic fumes, which smell like melting plastic and metal, are produced by overheating electronic circuits or components. Sadly, it usually means the life force of your component has gone on to join the great beyond. To prevent this while soldering, you can use a heat sink—a special spring-loaded clip or standard large-sized metal alligator clip that is placed between the joint to be soldered and the electronics. The metal clip will absorb the heat before it fries your circuit.

Soldering Linkbox

Make: Skill Set: Soldering (*http://bit.ly/ss-solder*)

Mini Fume Extractor (*http://bit.ly/mini-fume*)

Carnegie Mellon Lead Soldering Safety Guidelines (*http://bit.ly/1s7wgiN*)

U.S. Dept. of Energy Berkeley Lab Safe Soldering Work Practices (*http://bit.ly/safe-solder*)

Soldering Is Easy (free printable comic book) (*http://bit.ly/solder-comic*)

Adafruit Guide to Excellent Soldering (*http://bit.ly/ada-solder*)

SparkFun: How to Solder (*http://bit.ly/sf-solder*)

Learn to Solder (*http://bit.ly/learn-to-solder*) by Brian Jepson, Tyler Moskowite, and Gregory Hayes (2012, O'Reilly Media)

The Makerspace Workbench by Adam Kemp (2013, Maker Media)

Make: Electronics by Charles Platt (2009, Maker Media)

Directions
Step 1: List Your Requirements

This project is about developing the soldering skills while building a BEAM Miller Solar Engine, and then attaching it to a body designed from recycled or repurposed materials.

Step 2: Plan Your Project

If you have never soldered it would be helpful to practice on a learning kit or by joining pieces of wire before tackling this project. As mentioned above, you can choose to use the Solarbotics solar panel with a printed circuit on the back, which will make it easier to attach the components. The freeform option is challenging but doable for a novice. Be prepared to spend more time, and work patiently to get everything right. Using a helping hand tool to hold the components in place while you solder will make things a lot less stressful.

The standard Solar Wobblebot also uses a "windshield" made from the clear plastic top of a soft drink cup, with a hole for the straw at the top. If your motor is small enough to fit through the hole, you can mount your solar panel(s) on the top, as in the original design. If not, the solar panel(s) can be attached to the inside of the Solar Wobblebot. In that case, you can use a large clear container (like a takeout food container) as a windshield to cover the entire assembly.

Step 3: Stop, Review, and Get Feedback

You can get more information and see some different variations of freeform Miller Solar Engines on the Solarbotics website and in the book *JunkBots, BugBots & Bots on Wheels* by BEAM creator Mark W. Tilden and Solarbotics chief David Hrynkiw (see "Unevolved Robot Linkbox" on page 103).

Step 4: Build Your Prototype

Directions for Using the Solarbotics Circuit Board/Solar Panel:

 If you are not using the Solarbotics solar panel, skip ahead to the freeform instructions.

The directions below are adapted from the instructions that come with the Solarbotics solar panel with the circuit board on the back (Figure 3-8); I've made a few minor changes. (You can find a link to Solarbotics' instructions in "Unevolved Robot Linkbox" on page 103.)

Figure 3-8. *Solarbotics circuit board/solar panel and components for long-running Miller Solar Engine.*

1. Before you attach each component, use the wire clippers to trim the legs so they fit on the board without hanging over too much. They should be about 3/8 of an inch long (1 cm).

2. Hold the circuit board so the writing is right side up. For the following steps, as you attach each component, you will be going around the edge of the circuit board in a counterclockwise direction, starting at the middle of the top edge. It helps to apply a little solder before attaching the component to the board. You can either apply it directly to one lead (recommended by *Make:* editor Frank Teng), or to the metal pad on the circuit board (the Solarbotics method). The preapplied solder will help hold the component in place. Add a little more on top of the first leg as needed. Then feed solder to the second leg to attach it as well.

3. Take the PN2222 transistor and hold it flat side up. Solder the legs to the three oblong pads in the middle of the circuit board.

4. Take the 1381 voltage trigger and hold it rounded side up. Solder the legs to the three oblong pads to the left of the transistor.

5. Hold the disc capacitor so the positive side (marked with a +) is over the round pad to the left of the trigger and the negative side is over the square pad. Solder in place.

6. Take the diode and hold it so the end with the dark stripe is over the square pad on the left side of the circuit board, and the other end is over the round pad. Solder in place.

7. Take the barrel capacitor and hold it so the leg on the side with the dashed negative signs is over the square pad on the right side of the circuit board and the positive leg is over the round pad. Solder in place.

8. Take the leads from the motor. For extra stability, you can attach them to the circuit board "backwards"—so that the exposed metal tip of the wire is even with the edge of the circuit board and the insulated part goes back across the circuit board. Solder the motor leads onto the large round and hexagon-shaped pads on the right of the top edge. It doesn't matter which lead goes to which pad.

That's it! You're ready to skip ahead to Step 5, testing the solar engine and assembling the Wobblebot body.

Freeform Miller Solar Engine

 If you are using the Solarbotics solar panel circuit board, you can skip this section.

You will be using glue dots and solder to create a conga line of components that are connected to each other, starting with the PN2222 transistor.

For purposes of keeping left and right clear, consider all instructions to be from the point of view of holding them so the rounded back of the PN2222 transistor is towards you. When you need to bend a leg on a component, do it (gently) not more than

1/4 inch (6 mm) down from where it comes out of the plastic casing. All bends should be right angles unless otherwise noted.

1. Take the PN2222 transistor and gently bend the outer legs out to the sides.

2. Take the 1381 voltage trigger and hold it so the flat part is facing towards you. Bend the middle and right leg back, away from you. Then bend the left leg towards you.

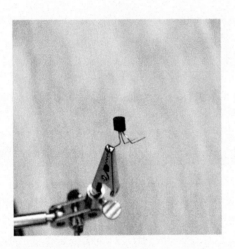

3. Use a glue dot to attach the flat side of the transistor to the flat side of the trigger. (You may want to mark the trigger with a white touch-up pen or bit of colored nail polish so you can tell them apart.) The left leg of the voltage trigger—which is pointing towards you at a right angle—should be touching the middle leg on the transistor. Solder those two legs together where they meet.

4. Use a glue dot to attach the disc capacitor to the back of the trigger, with the capacitor's legs on the outside of the bent legs of the transistor. The little positive sign on the yellow capacitor should be facing out. Solder the left leg of the capacitor to the middle leg of the transistor. Then solder the right legs together where they meet.

5. Take the diode and lay it along the right side of the three components, so it is touching (or resting on the bend of) the outermost legs on that side. Make sure the black-striped end is facing towards the transistor (and you). Solder the diode leg at the black-striped end

to the transistor leg where they meet. Solder the other leg to the trigger and the capacitor. (They are all connected.) Clip the excess wire at the end nearest the capacitor.

6. Take the barrel capacitor and hold it so that the negative side (with the negative signs running up the side like a dashed line) is on the right side as you're looking at it. Bend the right leg of the barrel capacitor out to the side and then forward, towards you.

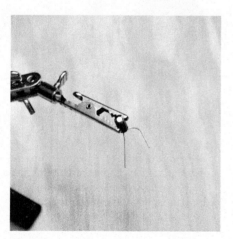

7. You can round off those bends a little, but leave some clearance so that leg doesn't touch the conga line when it is attached. Use a glue dot to connect the barrel capacitor to the disc capacitor. It should be positioned so that the left (positive) leg of the barrel capacitor is touching the middle leg of the trigger (which is already soldered to the left leg of the disk capacitor). Solder them together where they meet.

8. Position the right (negative) leg of the barrel capacitor so it touches the right leg of the transistor. Solder them together where they meet.

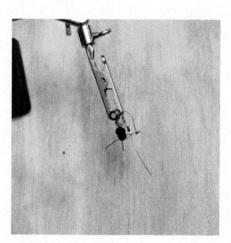

The Solar Panels

Follow these directions if you're making the freeform version. If not, skip to Step 5.

1. If you are using solar panels from solar garden lights, you need to connect two together to generate enough voltage for your Miller Solar Engine.

3. Cut the wires going to the solar cell as far from the cell as possible.

2. Open the solar garden lights and take out the battery.

4. Remove the solar cells if you can do so without damaging the wires. If not, use the wire cutter or other short clippers to trim the soft plastic around them, leaving a frame to hold the panel and the wires.

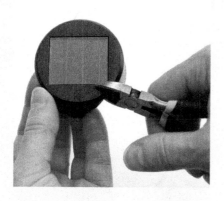

ly on the wire. Slip it over one of the wires on the outside edges of your joined solar cells and slide it down as far as it will go. Then—with the ends of the wires facing towards each other—twist the metal ends of the two outer wires (positive from one panel, negative from the other) together. Solder them together, being careful to keep the soldered joint narrow enough for the heat shrink tubing to fit over it. Slide the tubing over the joint. Use a heat gun (a crafting tool that's like a superhot hair dryer) to shrink the tubing until it fits snugly over the joint.

Alternately, wrap the joint tightly with a little electrical tape.

5. Then strip about 1/4 inch (6 mm) of insulation from the ends of the wires. The colors of the wires should indicate which is positive and which is negative (negative is usually black).

6. Hot glue your two solar cells together so that the positive wire from one is next to the negative wire from the other.

8. Take the remaining negative wire on your solar panels and twist it around the bent-out right leg of the barrel capacitor. Solder it on carefully. Take the positive wire from the solar panel and solder it to the positive leg of the barrel capacitor, leaving a little room for one more wire to be attached there.

7. If using heat shrink tubing, cut a short piece wide enough to just fit a little loose-

The Motor

Follow these directions if you are making the freeform version. Otherwise, skip to Step 5.

1. If you are using a recycled motor, you may want to test that it works with the solar engine before soldering it on. Try twisting the wires on or using alligator clips.

2. If using heat shrink tubing, cut two short pieces and slide them onto the motor wires, as far down as they will go.

3. The motor does have positive and negative leads, but it will work either way around. It will just spin in opposite directions, depending on how it is connected. Twist one of the motor wires onto the positive (left) side of the barrel capacitor. This is where you just attached the positive wire from the solar panel. Solder the motor wire on as well.

4. Solder the remaining motor wire to the left leg of the transistor. Slide the heat shrink tubing up to cover the solder joints and shrink to fit with the heat gun.

Step 5: Test Your Design

Before you attach it to the Solar Wobblebot body, test your Miller Solar Engine circuit.

1. You may want to make a little shoe for the shaft of your motor, especially if it's short. Remove the eraser from a standard pencil, squeezing the metal sleeve holding it on with your pliers to loosen it if needed. Push the shaft of the motor into the end of the eraser. Be sure to leave enough clearance to allow the shaft to spin. If you decided to skip the shoe, put a small piece of masking tape on the end of the motor to test it.

2. Before you test the solar panel, check to see if there is a protective sheet of clear plastic covering it, and if so, remove it. Expose the solar panel to bright sun or artificial light. In a minute or two, you should see the motor spin for several seconds. If it doesn't, check the Troubleshooting tips in Step 6 before going any further.

The Solar Wobblebot body is quick and easy to assemble.

To make sure everything fits, you may want to prototype it with glue dots. That lets you move things around as needed before you attach the pieces permanently with hot glue.

1. If you are using a cup dome with an opening in the top and your motor fits through the hole, gently push it through so that the solar panels are on the outside. Be careful not to tear the wires. Attach the solar panels over the opening.

2. Attach the motor to the CD with the shaft sticking down through the hole.

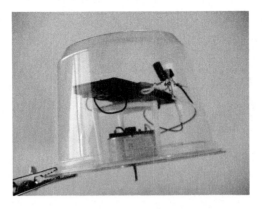

3. If the solar panels are going inside the Solar Wobblebot body, mount them on a pizza box spacer, upside-down ketchup cup, or other platform that is attached above the motor. Be careful with the components and wires as you position them.

4. Attach the windshield to the CD, fitting it carefully over the motor and other components. Put your pieces in the brightest area you can find to see if your Suped-Up Solar Wobblebot will spin and move. If everything works, go ahead and attach permanently with hot glue.

Step 6: Troubleshoot and Refine

If your Suped-Up Solar Wobblebot isn't running, check for broken wires, broken soldering connections, or shorts (where two metal pieces that shouldn't be touching are touching). Also make sure the shaft of the motor can spin freely.

If you are using a recycled motor and did not check to make sure it works with the solar engine, that may be your problem. Some small motors that run fine off of a battery will not work with a solar panel. You must have a low inertia motor that doesn't take a lot of current to get started. It's also possible that your motor doesn't have enough torque (circular force) to move the weight of the Wobblebot. In that case, you can try to convert it into a "flag-spinning" BEAM bot (see Step 7).

Step 7: Adaptations and Extensions

Once you've got your Miller Solar Engine working, you can make an ultra simple bot by simply turning whatever motor you use facing up. Instead of moving around, it can spin a small flag, sign, or colorful streamers attached to the shaft with a cocktail straw.

For more advanced DIY BEAM body designs, check the Instructables website and *make-zine.com*, or try one of the Make: kits from Solarbotics or MakerShed.

Unevolved Robot Linkbox

Junkbots, Bugbots, and Bots on Wheels companion website (*http://junkbots.solarbotics.com/*)

Original Bristlebot how-to (*http://bit.ly/emsl-bristlebot*)

Solarbotics retail site (*https://www.solarbotics.com*)

Miller Solar Engine kit documentation (*http://bit.ly/mse-solar-kit*)

Home Training Tools DC motor (*http://bit.ly/el-motor2*)

Absolute Beginner's Guide to Building Robots by (former *Make:* editorial director) Gareth Branwyn (2004, Que)

JunkBots, BugBots & Bots on Wheels by Mark W. Tilden and David Hrynkiw (2002, McGraw-Hill Osborne Media)

Robot Friends and Helpers 4.

I am C-3PO, human-cyborg relations. And this is my counterpart, R2-D2.

— C-3PO

Much of the research on robots is based on the idea that they can be much more than smart machines—they can also be our friends and helpers. So scientists are always looking for ways to make it easier for people and robots to understand each other. The fire hydrant–shaped character R2-D2 from Star Wars may only be able to speak in beeps and whistles, but fans will tell you the little droid is smart, loyal, and very cute. His companion C-3PO is much more humanoid in speech, appearance, and behavior, but he's also somewhat less cuddly. That's why the field of "social robotics" is investigating ways to smooth the interface between organic and mechanic beings.

The effort has several challenges. First, there's the robots' artificial intelligence, or AI—the way they think and behave. All robots think like machines. But robots that behave like machines are difficult to talk to. One line of thinking says that figuring out how to give robots a personality will help make them more people-friendly. In 1997, then-MIT grad student Cynthia Breazeal (who went on to found the MIT Media Lab's Personal Robots Group) created Kismet, the first social robot. Although Kismet didn't speak a human language, it could make appropriate sounds and gestures that responded to what a person said to it. Kismet's programming let it learn from conversations and interactions with humans.

There's also the issue of how robots look. Robots that resemble mechanical tools are efficient but hard to warm up to. So some scientists focus on giving robots a welcoming appearance. One way is to up the cute factor. Robots with child-like features—big eyes, an innocent expression—are easier to relate to. Breazeal at MIT created the robot Leonardo with the Stan Winston Studio, a Hollywood special effects company. Leonardo was a fur-covered animatronic creature that looked like a cross between a puppy and a human toddler. When you talked to Leo, it would search your face with its huge eyes, and express its feelings by pivoting its ears up or back and gesturing with its little arms. At Yale, the Social Robotics Lab uses robots like Pleo, an animatronic toy baby dinosaur, and Keepon, a yellow rubber snowman-shaped dancing robot whose only features are two eyes and a little button nose, to engage children who have trouble with social interactions.

Then there's the problem of how robots move. A robot designed to interact with humans needs to be compliant—able to adjust its speed and power to avoid causing injury to anyone around it. One industrial robot designed to address these needs is Baxter from ReThink Robotics (Figure 4-1). Baxter's friendly tablet face swings around to focus on you when you touch its arm. To teach it what to pick up and where to place it, you just move its lightweight arms to the right spot and push a few buttons. And it works at a human cadence, making it safer than high-speed machinery.

Figure 4-1. *The Baxter industrial and research robot scales its movements to human speed, so it's safer to work around. Credit: Rethink Robotics.*

In this chapter you'll design a chatbot—a computer program that may someday be the "brain" of a humanoid robot—and explore the limits of the *Uncanny Valley*, the spooky place where artificial creatures look too real to be believed.

Social Robotics Linkbox	
Cynthia Breazeal (*http://bit.ly/breazeal*)	Keepon (*http://beatbots.net*)
Nao from Aldebaran (*http://www.aldebaran.com/en*)	Yale Social Robotics Lab (*http://scazlab.yale.edu*)
	Baxter (*http://www.rethinkrobotics.com*)
Pleo (*http://www.pleoworld.com/*)	

Project: Make a Chatbot Program

Figure 4-2. *The finished Scratch Cat sprite*

What Is a Chatbot?

A chatbot is a computer program that can talk with people in natural-sounding language. It includes programs that communicate by reading and writing text, and those that can speak and understand verbal commands.

What It Does

Businesses use chatbots to steer callers to the proper extension and information. Websites use them as for customer assistance. They also act as a voice-activated interface for smartphones and other devices.

Where It Came From

Chatbots were originally designed as an intellectual exercise. In 1950, computer pioneer Alan Turing wrote an article for the philosophy journal *Mind* in which he pondered the question of whether machines think. One way to tell, he suggested, would be an experiment in which a person had to guess whether he was conversing with a computer or a human being. The "Turing Test" has come to be the standard for judging chatbots. In 1990, inventor Hugh Loebner created the Loebner Prize, an annual competition that offers $100,000 and a gold medal to the chatbot that can pass the Turing Test in spoken conversation. A silver medal prize of $25,000 is offered to any program that can fool half of the panel of judges into believing it is human based on a text message conversation. So far no chatbot has claimed the top prizes, but a bronze medal is given to the best competitor every year. In 2012, inspired by a paper that proposed a new Turing Test based on the way children learn to use language, a new Junior prize category was added that is judged by a panel of 12- to 14-year-olds.

The first actual chatbot, Eliza, was a text program developed by MIT professor Joseph Weizenbaum in 1966. Eliza was a take-off on the kind of therapy in which a psychologist turns every response from the patient into a question. For instance, if a user told Eliza they were feeling sad, the computer might respond "Why do you think you are feeling sad?" Eliza was so convincing that test subjects asked to be left alone with "her" so they could talk privately. In 1995, Richard Wallace created an open source program called Artificial Intelligence Markup Language (AIML) to design his chatbot ALICE, which stands for Artificial Linguistic Internet Computer Entity. AIML is still used today by chatbot researchers and hobbyists.

In 2013, the Loebner Prize bronze medal went to an online text chatbot named Mitsuku whose talents include "Yo Mama" battles with feisty fans. British musician Steve Worswick created Mitsuku as a way to entertain visitors to his site. Mitsuku also won the Funniest Computer Ever competition in 2014, a contest organized by computer science professor Sam Joseph of Hawaii Pacific University with the goal of finding a chatbot able to compete head-to-head with a human comedian.

Perhaps the most lifelike chatbot to date is Siri, the personal assistant built-in to Apple's iPhone and other products. Siri takes your verbal musings and uses them to control built-in apps like the alarm or music player and to do online searches when you want to find a movie playing near you or book a reservation at your favorite restaurant. But users also like the fact that Siri "knows" a lot about popular culture and can carry on interesting conversations. Siri's playful personality is so appealing that it inspired the 2013 movie *Her*, which revolves around a man who falls in love with the voice on his cell phone.

How It Works

To talk to humans, chatbots have to juggle a number of tasks at the same time. First, a chatbot that responds to spoken language must contain some kind of speech recognition software. After all, it can't give you an answer if it doesn't hear you correctly. Speech recognition software takes your spoken language and converts it into a digital form. It then analyzes each individual sound to figure out which phoneme (small units of sound that make up the words of a given language) you're using. English has about 40 phonemes; other languages have more or less. As it identifies the stream of phonemes, it compares them to statistical models to guess how they are most likely to be combined into words and sentences. It's a difficult task, but speech recognition software has come a long way in just a few years. Still, a personal assistant like Siri can have trouble deciphering accents that vary too much from the standard pronunciation, such as a Scottish brogue, until it "learns" the particular verbal patterns of its user. It does this by keeping track of when you like an answer and when you repeat a command to make sure you're understood.

No matter whether you're talking to your chatbot verbally or via text, however, a chatbot must be able to understand everyday conversation. In a computer language, there's usually only one right way to say something. Any little spelling or missing punctuation mark can throw off the entire program. But humans can express the same idea in countless ways, and new words and phrases pop up all the time. A chatbot needs to figure out exactly what you're referring to, using clues like context.

Unless a chatbot is just designed to keep you talking, like Eliza, it also needs a databank of information to draw upon. A chatbot like Siri can access both onboard and online information to find the answer you're looking for. And it can refine the accuracy of its answers through feedback from its own user and from the queries and responses of everyone else who uses the system, too.

Finally, chatbots must also know a little bit about human psychology. Eliza's open-ended responses were perfectly suited to encourage users to pour out their thoughts and feelings, as they would in a therapy session. But chatbots that can also keep you talking by learning how to push your buttons. In 1989, when the Internet was mainly used to send email and chat messages between colleges, a student at University College in Ireland named Mark Humphrys decided to test out an argumentative version of Eliza he called MGonz. He set up MGonz to receive his messages when he wasn't at his desk. An unknown student from Drake University in Iowa who tried to contact Humphrys after he had left for the night ended up trading insults with MGonz for over an hour. According to the transcript Humphrys found the next day, the Iowa student never realized it was a computer he was talking to, not a frustratingly rude computer science student.

What Happens When Two Chatbots Meet?

Figure 4-3. Cornell's "AI vs AI" video on YouTube

In 2011, Cornell University professor Hod Lipson set up two versions of an online chatbot called Cleverbot on adjacent computer screens and let them talk to each other (Figure 4-3). Cleverbot's repertoire of responses is completely based on past conversations with humans. Since it typically engages in around 10,000 conversations at once, this gives it an amazing variety of things to say. When Lipson posted the video on YouTube, it quickly went viral. Here's part of their conversation:

Shruti-bot: Hello there.

Alan-bot: Hiya.

Shruti-bot: How are you?

Alan-bot: Great. You?

Shruti-bot: I'm okay.

Alan-bot: That's good.

Shruti-bot: Are you good?

Alan-bot: Yes, I am good. I just said I was.

Shruti-bot: No, you didn't. I said I was.

Alan-bot: So you are a robot.

Shruti-bot: No, my name is Cleverbot.

Alan-bot: Yes, you are a robot and your name is Cleverbot.

Shruti-bot: I am a robot.

Alan-bot: Yes, I know.

Shruti-bot: Together, we are robots.

Alan-bot: I'm not a robot. I'm a unicorn.

Shruti-bot: But you said earlier that you are a robot.

Alan-bot: I did not.

Shruti-bot: I thought you did.

Alan-bot: You are mistaken, which is odd, since memory shouldn't be a problem for you.

Chatbot Linkbox

Chatbots.org (*http://www.chatbots.org*)

Loebner Prize (*http://www.loebner.net*)

Eliza simulation (*http://www.manifestation.com/neurotoys/eliza.php3*)

ALICE and AIML software (*http://www.alicebot.org*)

Mitsuku (*http://www.mitsuku.com*)

MGonz (*http://bit.ly/1mefNrK*) (Warning—not suitable for children!)

Cleverbot (*http://www.cleverbot.com*)

Cornell Cleverbot conversation (*http://creativemachines.cornell.edu/AI-vs-AI*)

Making the Project

You will be creating a text-based chatbot in an open source programming language called Scratch (Figure 4-4). Scratch's free drag-and-drop software was designed by the Lifelong Kindergarten group at the MIT Media Lab to help children learn the basic concepts of programming. Even though it's for kids, it's great for beginning adults, too—and it's fun! You can create your program right on the MIT-based website, and explore other people's projects as well. Copying and building upon other people's projects, or "remixing," is encouraged. Despite its simplicity, you can create quite sophisticated projects in Scratch.

 Scratch 2.0, the current version of the program, can be used online on the Scratch website. At the time of this writing, only Scratch 2.0 Offline Editor and Scratch 1.4 (the previous version) are available for download.

You will not be building a physical robot for this project, although you have the option to change the look of your chatbot's onscreen avatar.

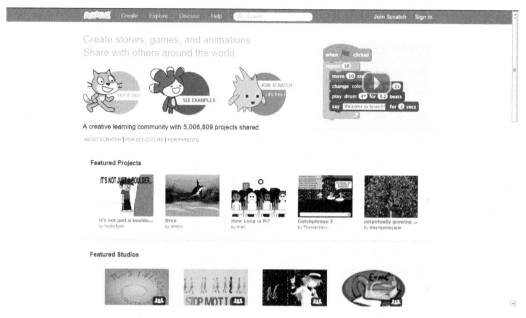

Figure 4-4. *The Scratch home page.*

Scratch and Other Graphical Programs

The beauty of graphical programs like Scratch is that instead of having to type in commands (and worrying about whether they are correct to the letter), all you need to do is move around "blocks" of code on the screen with the cursor. Blocks that can be used in sequence will snap together; try to put the wrong type of command in the wrong place and it just won't fit. Of course, you still need to figure out which commands to use and get them in the right order, but drag-and-drop programming languages make it simple to switch blocks around and try out new programs over and over.

Drag-and-drop software is seen as a powerful education tool. By lowering the barrier to creating a first program, it attracts students who don't necessarily consider themselves tech-minded. Carnegie Mellon's free program Alice—created by Randy Pausch (the late computer science professor whose "Last Lecture" became a viral video and book) is another drag-and-drop program that lets users create animations and tell stories. Its graphics were designed by computer game giant Electronic Arts, Inc. (EA). Alice has been a stepping stone for students going on to learn traditional programming languages like Java.

Graphical software has also been used to program robots, microcontrollers like the Arduino, and other physical computing devices. Lego has long used the drag-and-drop program LabVIEW to control its robotics kits. The downloadable Scratch 1.4 can be used with the Lego Education WeDo kit for elementary school–aged children (with Scratch 2 to be available in the future). It can also be used to control the Xbox Kinect and with MaKey MaKey, a kit that turns any kind of conductive object—like bananas—into a keyboard. Scratch for Arduino, developed by the Citilab Smalltalk Team in Spain in 2010, is a modified version of Scratch that can interact with Arduino boards. A similar program, the open source Ardublock, was designed by David Li of the Shanghai makerspace XinCheJian.

Don't forget to document your work!

Project Parameters

- Time Needed: 2-3 hours
- Cost: None (computer and Internet access needed)
- Difficulty: Easy to moderate
- Safety Issues: None

What You Need to Know

- Skills You Already Have: Computer use (logging onto websites, saving work in progress)
- Skills You Will Learn: Programming concepts, Scratch software

Gather Your Materials

- Computer and access to the Internet software or online tool

Directions
Step 1: List Your Requirements

At a minimum, your chatbot should be able to carry on a short conversation with a human in natural-sounding language. For your first try at programming, that should be enough. Once you've got the hang of it, though, you can add features like a personality—funny, thoughtful, obnoxious—or give your chatbot a task, such as playing Twenty Questions or (pretending to) tell the future.

Step 2: Plan Your Project

When you write your Scratch program—or "script," as it's called in Scratch—you will actually be creating a little interactive animation, starring a character or object called a "sprite." Each sprite does whatever you tell it to do—moves, dances, talks (using text balloons), or make sounds (using audio files).The default Scratch sprite is the Scratch Cat, the orange cartoon character that serves as the Scratch mascot and logo. You can also choose a sprite and background from Scratch's library of available images, or upload your own. Scratch has its own online graphics software that lets you alter any image or draw your own from scratch (no pun intended).

Before you get started, take a few moments to go through the built-in Scratch tutorial (Figure 4-5). Go to the Scratch home page (*http://scratch.mit.edu*) and click "Try It Out" or "Create." A blank editing page will open up. The step-by-step animated tutorial should be open on the right side of the screen. If you need to refer to it again at any point, just click on the little question mark in the top righthand corner.

Figure 4-5. *The Scratch tutorial.*

The tutorial will give you a quick idea of what tools are available and where to find them. But there's still a lot to learn about using Scratch *and* about writing a chatbot program. As with many kinds of programming languages and programs, the easiest way to learn how they work is to copy someone else's project and then tweak it to see what happens. In Scratch, the way to do that is to open up any project page and click See Inside.

To get you started, I've created a very basic Step by Step Chatbot on the Scratch website that you can use as a model for your own Scratch program. It incorporates a few programming concepts like nesting, loops, and "if" statements (more on those soon). To see how it works, go to the project page for Step by Step Chatbot (*http://scratch.mit.edu/projects/19361862*) and try it out. Enter different answers to see how it affects what happens. Then click on See Inside and take a look at the program itself. Don't worry if it doesn't make sense at this point—you'll go through the process of putting it together step by step soon. As with all projects on the Scratch website, you can remix my Step by Step Chatbot project and use it as the basis for your own chatbot.

As already mentioned, you can download Scratch to your computer and work offline, but if possible use the online version. Sign up for a (free) Scratch account and you can save your project and share it with the online learning community as well. All it takes is a username, password, and an email address. However, even if you don't sign up, you can still create a project while you're online, access the tutorials, and check out the code behind other projects.

Scratch Cheat Sheet

Some quick tips to help you get started.

The Scratch editing page has three columns.

On the left is the Stage where your animation will play. Underneath is the toolbar for the backdrop and for sprites. You can customize them by clicking the icons in the toolbar directly above it. Above the Stage is a blank field where you can enter a name for your project, and above that is the toolbar for the page.

In the center of the page is the Block Palette. It has the command blocks you need to build your project, arranged by category. The category names are not quite intuitive, but they are color-coded, so if you see a command similar to the one you're looking for you can try the same category.

To to the right is the Scripts Area. To add a command to a script, just click on the block you want to use and drag it over to the Scripts Area.

To add a second command, drag it over to the Scripts Area and nudge it under the first block until it snaps into place. Continue to add commands the same way.

To move a bunch of connected blocks, click on the top block.

To disconnect part of a stack of blocks, click on the topmost block of the ones you want to move and pull them down and away from the rest.

You can change the number or text in the white field of a command by simply typing right over it.

You can layer certain blocks on top of one another. For instance, some blocks have pointy-sided blank spaces on them. Pointy-sided commands or operators (such as "=") will fit into those spaces. Just drag one block over the other until it clicks in. Once you start building multiple layers, double-check that you are plugging the right block into the right space.

Scratch lets you create a loop, a bit of programming that will repeat over and over until a certain condition (which you specify) is met. The framework of a loop is a block from the tan Control category of the Block Palette with a mouth-like opening. The mouth will stretch around whatever size stack of blocks you place inside it. You can also put one loop inside another—that is called "nesting." Think of those wooden Russian dolls of diminishing size that nest one inside the other.

You can check your program at any time by clicking on the stack of blocks and watching them play on the Stage.

Step 3: Stop, Review, and Get Feedback

The Scratch website has tons of support, including the above-mentioned tutorials, forums where you can ask questions, and materials for educators and parents. There are also some books that are useful, which I have listed in "Graphical Programming Linkbox" on page 123. One handy teacher-created resource is the Scratch Chat Bot Assignment worksheet (*http://bit.ly/chat-assign*).

Step 4: Build Your Prototype

These directions show you how to put together the Step by Step Chatbot. Be sure to check them against the illustrations.

Brackets like this [...] indicate a field that gets filled in by you when writing the program, or by the user when running the program.

1. Go to *scratch.mit.edu*, sign in, and click Create. For now it's probably easiest to use the default Scratch Cat (the orange Scratch logo). Click File and save your project. This will preserve at least some of your work if you close the page accidentally.

2. Start the Step by Step Chatbot script by going to the purple Looks category of the Block Palette and dragging the block "say [Hello!] for [2] secs" over to the Scripts Area. Click on the block in the Scripts Area to watch it play.

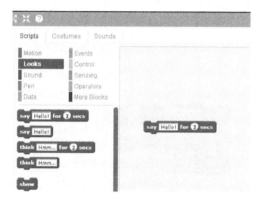

3. Now for a bit of computer logic! You will give the program some data and tell it what to do with it. Go to the light-blue Sensing category of the Block Palette, and drag and drop the "ask [What's Your Name?] and wait" block to the Scripts Area. Nudge it under the first block until it snaps into place. Click on the connected blocks to play that bit of code. On the Stage, you will see a field open up where users can enter text. That text—in this case, the name —will then become data which is stored in an "answer" block. The program can then use the data in various ways.

4. To make the Scratch Cat say "Hello, [name]!" you have to layer several blocks on top of each other. Go back to the Looks category of the Block Palette and add another "say [Hello!] for [2] secs" block to the bottom of the stack of commands.

5. Then go to the green Operators category of the Block Palette and grab an oval-shaped "join [hello][world]" block. The "join" block lets you fit two things into one space. Drag the green oval over to the Scripts Area, and hover it over the purple "Say" block you just added until you see the white field begin to glow. The green oval should then snap into place in the field. Grab a second "join" block and drag it on top of the first "join" block. Snap it into the second white field, the one that says "world." Now you have three fields to play around with. In the first, capitalize "hello" and add a comma and a space ("Hello, "). In the last field, replace the word "world" with an exclamation mark (!). In the remaining white field (which says "hello"), you'll put an "answer" block that will contain the name the user gives you. If you want the user's entry to be shown on the Stage as the program runs, check the box next to the "answer" block in the Sensing category of the Block Palette.

6. Open the light-blue Sensing category of the Block Palette and drag an oval "answer" block over to the middle field until it snaps in. Tip: Make sure only the field where you want to insert the "answer" block is lit up, not the entire "join" block. You may knock the blocks you already assembled out of place! Click on the stack of blocks and see how your program runs. You should see the Scratch Cat ask you what your name is, and then greet you by that name.

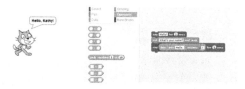

7. Add another "ask" block to the bottom of the stack, and change the text to "Do you want to chat?"

8. Next up is a loop. The loop for the Step by Step Chatbot will be a "repeat until […]" block. Each time the loop plays all the way through, the program will ask whether the user wants to keep chatting. As long as the answer isn't "No," the program will go back to the top of the loop and start over. When the answer is "No," the program will leave the loop and continue on. Go to the tan Control category of the Block Palette and drag a "repeat until […]" block over to the Scripts Area, but don't connect it to the stack of commands yet. You'll build and test this bit of code separately, to make it easier to work with. The condition needed to end the loop is for "answer" to equal "No."

9. In the green Operators category of the Block Palette, you can find a pointy-ended block with two blank fields and an equals sign ("=") in between . Drag it over to the empty field on the "repeat until" block and snap it in. Then go to the light-blue Sensing category of the Block Palette and drag an "answer" block over to the first empty field of the "equals" block. Type "No" in the second empty field.

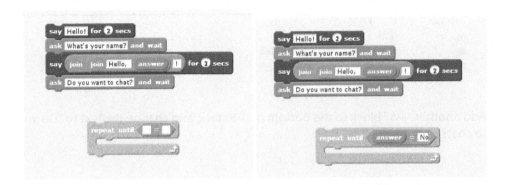

10. Now get a little conversation going back and forth between the chatbot and the person using it.

- From the light-blue Sensing category of the Block Palette, drag an "ask and wait" block into the mouth of the loop. Change the question to, "OK, what would you like to chat about?"

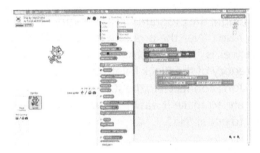

- Drag a second "ask and wait" block underneath the first. Change this question to "Do you like [answer] a lot, a little, or not at all?" To do that, you'll need to layer green "join" blocks and drag over an "answer" block the same way you did in #3 above. Don't forget to include spaces in the text that comes before and after the "answer" block. (By the way, when computer programmers want to avoid writing out the same set of instructions in more than one place, they tell the computer the equivalent of "do what you did in #3." That's called a *subroutine*.) Run this stack to see if it's working before going on.

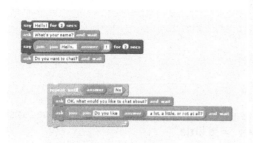 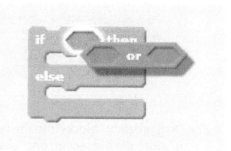

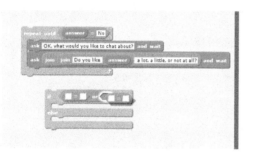

11. To analyze the user's answer so the program knows what to do next, you'll use another kind of Control block: the "if-then-else" block. You may remember if/then statements from high school geometry. They work exactly the same way here. *If* a certain condition is met, *then* the program takes one course of action—or *else* it takes another action. Since you asked the user to choose from among three responses, you'll need to give the program directions for all three. In this case, the answers "a lot" and "a little" both lead to the same course of action. Here's how to assemble that bit of code:

- Drag an "if-then-else" block to the Scripts Area from the tan Control category of the Block Palette. Don't connect it to any other stacks of blocks yet.

- Drag a green "or" block from the Operators category of the Block Palette and plug it into the pointy-ended hole between "if" and "then."

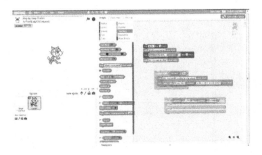

- Drag green "=" blocks into the holes in the "or" block. Make the first one "answer" = "a lot" and the second one "answer" = "a little."

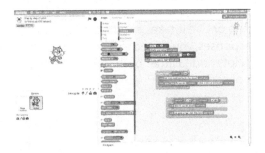

- Drag light-blue "ask and wait" blocks into each mouth of the "if-then-else" block. Type "What do you like most?" into the first and "I'm sorry to hear that. Why not?" into the second.

- Drag the "if-then-else" block into the mouth of the "repeat until" block and plug it in under the two "ask and wait" blocks. You can now test the loop by clicking on it to see how it runs.

12. Feel like drawing out the conversation? Add another leading question that will keep the person using your chatbot program talking. For instance, you can drag an "ask and wait" block under the "if-then-else" block, making sure it is still inside the mouth of the "repeat until" loop. Then type a line of dialogue such as "Really! Why do you think that is?"

13. Time to give the loop something to judge. The last piece of dialogue in your loop should be a yes or no question that will signal whether or not the person using it is ready to stop. Drag an "ask and wait" block below the last one (still inside the "repeat until" loop). Insert a comment such as "I totally agree! Is there anything else you'd like to talk about?" If the answer is "No," the program will exit the loop. Any other answer will signal the loop to repeat again.

14. End the conversation with a little good-bye message. Drag a purple "say [Hello!] for [2] secs" block to the Scripts Area and attach underneath the loop. In the text field, type in "OK, thanks for dropping by!"

15. Drag the entire stack of code over to the first segment you wrote and attach them. As a final touch, add a "when clicked" block from the brown Events category of the Block Palette to the very top. This tells the program to run when a user clicks the green flag on the project page. You can add a tan "stop [this script]" Control block at the end, but it's not really needed in this program. Your chatbot is now finished and ready for its first tryout!

Break the Code: "Hello World"

When you've just put together a new electronics device or started working in a new computer language, the "Hello World" program is the first thing you try to make sure it's working right. The phrase is generic: it doesn't have to literally make the machine say "Hello World."

Step 5: Test Your Design

Test your chatbot a few times yourself by running through the project, entering different kinds of answers. If you need to fix a line or want to add a step, this is the time to do it. Then invite friends to try it out. If you create your project on scratch.mit.edu and set it to Share, they can access it from their own computers.

Step 6: Troubleshoot and Refine

It can take a little while to get the hang of dragging and dropping the blocks just where you want them. See "Scratch Cheat Sheet" on page 114 for help.

As for refining your program, that's up to you. Once you've got a basic chatbot working, you can add as many levels of complication as you like. For instance, to keep things simple, the Step by Step Chatbot requires the user to type "No" to end the program—but that's not spelled out in the dialogue boxes. There are a number of ways to solve that dilemma—including accepting more types of negative answers, or getting more clever with the conversation. Take a look at the code for more advanced chatbots like the GLaDOS chatbot on the Scratch website (see link below) for inspiration.

Step 7: Adaptations and Extensions

This project can be adapted for use without a computer by simply writing it out on a piece of paper as a logic tree—kind of like those old *Choose Your Own Adventure* stories. Start at the top of the page with the first line of dialogue, and then branch off below each time the user gets to make a choice one way or the other.

You can also extend the project by giving your chatbot a physical body. Load it onto a phone or tablet (or a Raspberry Pi) and build a robot body to hold it.

Graphical Programming Linkbox

MIT Scratch website (*http://scratch.mit.edu*)

Scratch 2.0 beta download (*http://scratch.mit.edu/scratch2download*)

Scratch 1.4 download (*http://scratch.mit.edu/scratch_1.4*)

Carnegie Mellon's Alice (*http://www.alice.org*)

Scratch: How to Connect to the Physical World (*http://wiki.scratch.mit.edu/wiki/How_to_Connect_to_the_Physical_World*)

Scratch for Arduino (*http://s4a.cat*)

Ardublock (*http://blog.ardublock.com*)

Step by Step Chatbot (*http://scratch.mit.edu/projects/19361862*)

Scratch GLaDOS chatbot (*http://scratch.mit.edu/projects/17155051*)

And in print…

- *Learn How to Program with Scratch* by Majed Marji (No Starch Press, 2014)

Project: Picture Yourself in the Uncanny Valley

What Is the Uncanny Valley?

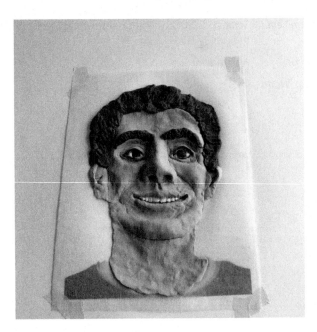

Figure 4-6. *An Uncanny Valley face.*

The Uncanny Valley is a theory that says people feel more comfortable with robots that look more human—but at a certain point, an artificial being that's just a little bit off becomes frightening or repulsive.

What It Does

Fear of falling into the Uncanny Valley makes it hard for designers in the fields of robotics and computer animation to make realistic-looking representations of humans.

Where It Came From

The Uncanny Valley was first described by Japanese roboticist Masahiro Mori in 1970. He suggested that a line graphing the steady increase in how much people liked humanoid robots as they became more realistic would take a sharp dip right before reaching the point where they were indistinguishable from actual human beings. At a robotics conference in 2013, Mori said he was inspired by his work on prosthetic limbs. He noticed that people would respond differently to an artificial hand that looked like a tool or machine as opposed to one that attempted to look like a real hand.

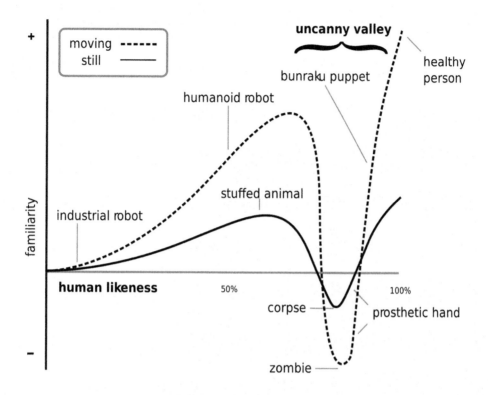

Figure 4-7. *A simplified version of Mori's Uncanny Valley diagram. Credit: Karl MacDorman under a GNU Free Documentation License.*

Of course, at the time Mori first developed the Uncanny Valley concept, robots weren't very realistic looking. So it took another 30 years for the idea to seep into popular culture. One inkling came in 2001, when children in test audiences for the 3D animated film *Shrek* began to cry at the appearance of a quasi-realistic Princess Fiona. That same year, a film called *Final Fantasy: The Spirits Within*, based on a video game, bombed at the box office in part because its computer-generated characters gave off a weird vibe. The 2004 movie *Polar Express*, which featured a motion-capture animated version of actor Tom Hanks, gave the term even more exposure. Today the idea of the Uncanny Valley is a factor that designers keep in mind when working on video games, computer graphics, 3D animation, and plastic surgery.

Robots balancing on the lip of the Uncanny Valley affect are most common in Japan. In 2011, Showa, Waseda, and Kogakuin Universities worked with the robot maker Tmsuk to create a simulated dental patient for the Yoshida Dental Manufacturing company. Hanako 2 had soft silicone skin, a tongue that moved, eyes that blinked, and a head that turned in response to a painful touch by the dental student. And roboticist Hiroshi Ishiguro of Osaka University in Japan made a robotic clone of himself named Geminoid. A former painter, Ishiguro started out building robots that looked more like trash cans or insects, but got a mixed response. So he decided to see how people would react to a nearly human robot (Figure 4-8). His silicone-skinned robotic doubles often wear the same serious expression as their maker. To add to the realism, he even tops them with hair from his own scalp.

Figure 4-8. *Hiroshi Ishiguro developed Geminoid HI-4 in his own image to explore whether human presence could be transferred to a remote location using robots. Credit: Hiroshi Ishiguro, Osaka University.*

Meanwhile, in the United States, David Hanson of Hanson Robotics creates robots that both look and act like real human beings. Hanson's android version of the science fiction author Philip K. Dick (whose book *Do Androids Dream of Electric Sheep* became the movie *Blade Runner*, featuring robots that don't realize they aren't real people) can autonomously answer an interviewer's questions while making appropriate facial gestures. Hanson has said he's constantly refining his Philip K. Dick android, and he'd love to see it star in its own television talk show.

How It Works

No one is sure how the Uncanny Valley works. In fact, there is disagreement about whether it really exists at all. In 2005, David Hanson wrote a paper disputing the Uncanny Valley, claiming that test subjects had no problem with his lifelike robots. A year later, Karl Mac-Dorman, professor of Human-Computer Interaction at the Indiana University Department of Informatics, compared the reaction of observers to a video clip of a stylized robot morphing into Hanson's Philip K. Dick robot and then into a photo of the actual author. His results suggested that other components, such as how strange or familiar a robot looked, also played a role in determining how much of an Uncanny Valley effect was felt.

But other studies seem to show a biological component to the reaction. A study done by neuroscientist Asif Ghazanfar at Princeton University in 2009 found that monkeys would avoid looking at photos that were close to but not exactly monkeys, compared with photographs or symbolic drawings. That may mean that the Uncanny Valley came about through evolution—perhaps because it alerted our ancestors that something was "off" about an individual whose appearance was different.

Motion may be another contributing factor. The body of the headless Big Dog robot from Boston Dynamics looks more like a keg of beer than a Doberman. But its legs run in a very dog-like fashion, which can be unnerving to watch. Likewise, passive-dynamic walkers—robotic sets of human-like legs that many labs study to determine how gravity can help make robots walk—are spooky to watch as they casually stroll, bodiless, down an incline. And on a commonly used chart of the Uncanny Valley, zombies ("the living dead") rank considerably lower than an ordinary corpse.

Hiroshi Ishiguro's goal is to create a robot that could pass the visual equivalent of a Turing Test and fool a human into thinking it is real. But Masahiro Mori himself has said he believes a better solution to the Uncanny Valley problem is to work on friendly designs that are not quite so real.

Making the Project

For this project you will explore the limits of the Uncanny Valley by making as close a likeness of yourself as you can. Will the result be cool or creepy? Survey your friends to find out.

 Don't forget to document your work!

Project Parameters

- Time Needed: 1–2 hours
- Cost: $10 or less
- Difficulty: Easy
- Safety Issues: None

What You Need to Know

- Skills You Already Have: Playing with modeling compound
- Skills You Will Learn: Making a lifelike model

Gather Your Materials

- Two copies of a black and white photo of your head, facing forward, against a white background, as large as possible (standard copy paper is fine)
- Waxed paper or tracing paper, about the size of copy paper
- Masking tape or glue stick or spray glue
- Poster board or other smooth stiff board
- Crayola Model Magic or other modeling compound, in skin tones or white
- Washable markers for coloring the modeling compound
- Clay modeling tools or kitchen utensils, including:
  — Unsharpened pencil with clean eraser
  — Toothpicks
  — Craft sticks
  — Plastic forks, knives, and spoons

Directions
Step 1: List Your Requirements

The goal of this project is to explore the Uncanny Valley by making a lifelike copy of your own face. To add to the realism, you will make a bas-relief model using modeling compound to raise the design a little from the background and create lights and shadows. You should use a stiff lightweight base if you intend to display it at an upright angle.

Step 2: Plan Your Project

The key to creating a model that looks as lifelike as possible is to start with a good photo. The more prominent the outlines of each feature, the easier it will be to extend them into the third dimension. Think about what other details will help your clone look eerily believable, and try to include as many as you can. For instance, individual teeth inside a mouth can look "too real," so try taking your photo with your mouth open or a big wide smile. But any wrinkle, scar or other imperfection will help move your model from artistic to hyperrealistic.

Like roboticist Hiroshi Ishiguro, you can also make your Uncanny Valley face seem more real by giving your double your own personality and emotions. Or instead of a friendly smile, use an exaggerated expression, like a robot that doesn't quite get social norms for human interactions. If the thought of building your own clone is too unsettling, try using someone you know in real life as a model, so you can tell what their features look like in three dimensions. Failing that, try to use photos from at least two views, from the front and from the side.

Step 3: Stop, Review, and Get Feedback

For ideas on how to take a robot face from cute to creepy, check out photorealistic artists whose paintings and sculptures give viewers those same uneasy feelings as humanoid robots. For instance, in 2013 sculptor Tony Matelli's life-size, full-color statue of a man sleepwalking in his underwear upset students when it was placed on the grounds of Wellesley College in Massachusetts. The man's zombie-like pose, bald head, pale skin, pot belly, and droopy drawers (as well as the fact that the nearly naked figure was situated on the grounds of an all-women's college) heightened viewers' feelings of disgust and fear.

Tips on using Model Magic or similar modeling compounds: Crayola Model Magic feels "foamy" and is easy to mold. It air dries quickly and becomes permanent overnight, but retains its rubbery appearance. Buy the smallest pouch or container you can, as once it is opened it should be used within a few days. Store opened packages in an airtight container, like a ziptop bag. As you work, use the smallest amount possible. It's easier to add more than to take away extra. One secret to making your model look real is to build it up in layers, letting them overlap, and then using your carving tools to create a little overhang. So the teeth should overhang the inside of the mouth, and the lips should overhang the teeth. That will create shadows that give your sculpture the illusion of depth.

Step 4: Build Your Prototype

1. If you want to make a model you can display, attach the photo you are working from directly to the board you are using for a base with a glue stick or spray glue. Otherwise, you can just tack it down with small pieces of masking tape at the corners. You may want to attach a sheet of waxed paper or tracing paper over your photo with masking tape that you can lift to check how closely you are following the guide photo underneath the modeling compound.

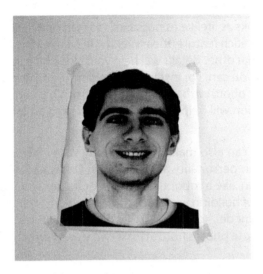

2. Model Magic offers a set of natural colors that work well for skin tones, but you can mix your own custom shades with washable markers. Try to mix up as much as you will need all at once, to keep the color consistent. To tint white Model Magic, take a hunk and flatten it somewhat into a thick pancake. Then take the tip of a washable marker and roll it back and forth on the pancake's top. When you've got about as much pigment as you think you'll need, start smushing and kneading the compound until the colors are evenly blended.

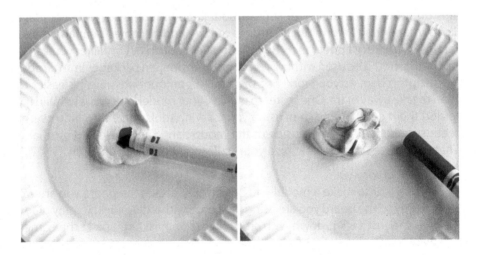

3. Start building the face by working on the highest parts first—the nose, chin, cheeks, and brow. Use a plastic spoon and other tools for smoothing the skin and digging creases where they appear on the photo. Remember that creating "overhangs" in places like the nostrils adds realistic shadows. And pay attention to the

shape of the nostrils, because they are unique for each person and add to the realism. Refer to the second copy of the photo as a guide.

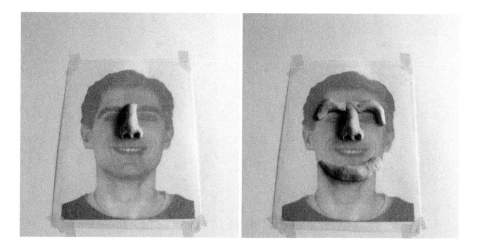

4. Next work on lower spots like the throat, mouth, and forehead. Build up the face piece by piece, connecting them as you go. For the ears, start with the highest parts, then fill in the flats. Again, the curves of the ear are unique, and the opening to the ear canal is another place to create an overhang that makes the model look realistic. Fill in any other skin areas, leaving the mouth and eyes open.

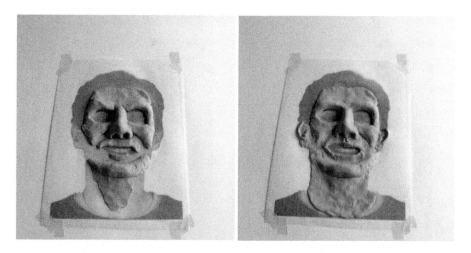

5. If the mouth in the photo is open enough to see the tongue or gums, mold that first. Make each tooth individually with white Model Magic, using your tools to shape them and keep them separate. For the eyes, start by making a flattened ball for the iris. Add little bits of white a little lower, tapering down at the corners to indicate the curvature

of the eyeball. You can use a black marker to draw on the pupils. Then add the eyelids and eyelashes.

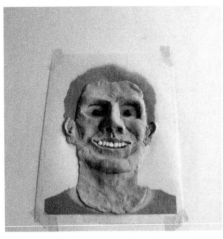 

6. With your lip color, add lips overlapping the teeth. Be careful not to mix the white with the surrounding skin color.

7. Making believable hair from Model Magic is difficult. For eyebrows, eyelashes, as well as hair on the scalp, indicate individual strands by using a fork as a carving tool. Add hair to the scalp in clumps, to give it dimension. Alternately, you can let your model dry overnight and then glue on artificial or real hair.

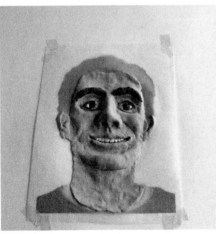 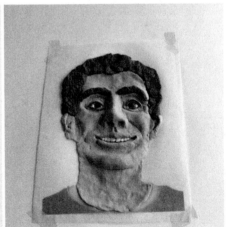

Step 5: Test Your Design

To see how successful you were with your model, take a survey of your friends. Ask them to rate it for factors such as how recognizable it is and how realistic. To measure the Uncanny Valley effect, also ask how it makes them feel—are they happy to see it, or does it make them feel weird?

Step 6: Troubleshoot and Refine

Unfortunately, if you're working with Model Magic, you basically will only get one shot to get it right. The compound is easy to work with, but it dries quickly. You'll need to complete your model in one session.

Be careful while you're working not to smush up a part you've already finished. Be especially aware of areas where different colors meet and try to keep them from running into each other.

Because Model Magic has a rubbery texture, it will add realism to your model's skin. But some areas, like the eyes and teeth, will look more believable if they appear slightly moist and shiny. You can highlight those areas by brushing on some Model Magic Glossy Glaze or other shiny finish coat when the modeling compound has dried. You can also dress up your model with artificial hair, or give it accessories such as glasses or earrings.

Step 7: Adaptations and Extensions

If creating a 3D version of your face from a flat surface is too hard, you can use a three-dimensional mask (art retailers have blank ones) or head form like those used to model wigs and build it up with the Model Magic.

You can also make the project simpler by keeping it two-dimensional and just tracing your photo onto tracing paper. Even a 2D line drawing can approach the Uncanny Valley if you include enough details. Or just focus on one feature, such as an eye or a mouth, that can in itself evoke an Uncanny Valley reaction.

As an extension, use your Model Magic doppelganger (or your actual face) as a mold to make a silicone mask. The Smooth-On company has kits and friendly tech support people to help you get started.

Or go digital and use panoramic photos and CAD software to make a 3D computer model of your face. You can then turn it into a pattern to build with cardboard or print it out on a 3D printer.

Uncanny Valley Linkbox

Masahiro Mori's Uncanny Valley paper (*http://www.androidscience.com/theuncannyvalley/proceedings2005/uncannyvalley.html*) "Uncanny Valley Revisited" robotics summit (*http://bit.ly/1edBKFc*)

Hanako 2 dental robot (*http://www.tmsuk.co.jp/english*)

Hanson Robotics (*http://www.hansonrobotics.com*)

Hiroshi Ishiguro Laboratories (*http://www.geminoid.jp/en/index.html*)

Crayola Model Magic (*http://www.crayola.com/products/all/model-magic/*)

Smooth-On silicone lifecasting materials (*http://www.smooth-on.com*)

Fun, Artsy Robots | 5

The world is your construction kit.

— Jay Silver, electronic toy inventor

One of the advantages of using robotic devices, with their machine bodies and computer brains, to do tasks for us is that robots can be so much more precise than humans. But robots can also be playful and creative. Toy and game designers use robotics to build movement and interactivity into everything from remote control quadcopters to talking dolls to electronic books that blur the lines between stories and games. Robots that make their own art are a kind of filter through which everyday things get a new interpretation. And of course, many robots are stylish enough to be considered works of art in themselves. For hobbyists, artists, and inventors, robots open up all kinds of possibilities.

One example of the overlap between robotics and art is Colour Chaser, designed by Japanese-born, London-based artist Yuri Suzuki. Colour Chaser is a small white plastic box with wheels containing line-following and color-sensing electronics. At a 2013 installation in Luxembourg, visitors were invited to use a black marker to draw a line on a sheet of paper for the bot to follow, then cover the path with scribbles of color. As the Colour Chaser passed over the different colors, it translated the RGB (red-green-blue) data into musical tones. Other artistic robots can even improvise. Artist Matthias Dörfelt's Robo Faber is a dinner-plate-sized shiny metal dome that drives around a sheet of paper on the floor, creating its own original doodles. Inside are two position-sensing electronics, an Arduino controller, and a permanent marker. By connecting randomized shapes it produces drawings of weird little creatures. Perhaps the best part is the way Robo Faber pauses as if thinking about what to draw next.

Today, electronics that respond to input much the same way as robotic devices do are also making their way into the world of toys and crafts. For example, Drawdio is a pencil that plays tones controlled by what you draw with it. It was developed by Jay Silver, then a graduate student in the Lifelong Kindergarten group at the MIT Media Lab, during a 2008 stint working with a school in the slums of Bangalore, India. Silver bought an electronic "harmonium" kit at a street market, sawed the keyboard off, and created a toy that would produce a sound by touching its wires to something conductive, completing the circuit as the keys on the keyboard had done. A friend pointed out that the graphite in a pencil could also be used to make a sound-producing circuit, which led to the idea of an electronic pencil that could "draw" music. A similar idea is behind MaKey MaKey, an electronic toy invented in 2011 by Silver and fellow MIT Media Lab student Eric Rosenbaum. (Both also helped develop the Scratch visual computer programming language.) MaKey MaKey is an Arduino-based printed circuit with six hookup pads. Using alligator clip wires, the MaKey MaKey can be connected to anything conductive, which can then be used like keys on a keyboard or a computer mouse. Bananas can become piano keys, and penciled lines on a sheet of paper can be used to play computer games.

Not surprisingly, hobbyists also use crafts materials to build actual robots and robot prototypes. "RobotGrrl" Erin Kennedy of Canada has won several awards at Maker Faires around North America with her RoboBrrd, an open source educational robot. You can buy a RoboBrrd kit or create a 3D-printed or laser-cut body from the online files—but the original RoboBrrd tutorial on Instructables was made using popsicle sticks, felt, and pipe cleaners. The little desktop toy has flapping wings, light-up eyes, and a beak that opens and closes. It's actuated by several servos and controlled by an Arduino shield, but the frame and skin are all crafts materials. Even the complicated hinges were made of coffee stirrers and twist ties.

In this chapter you'll learn to build a littleBits Plotter and an Arduino FiberBot that incorporates a programmable LED screen.

Artsy Robot Linkbox

Yuri Suzuki's Colour Chaser (*http://yurisuzuki.com/works/colour-chaser*)

Matthias Dšrfelt's Robo Faber (*http://www.mokafolio.de/works/Mechanical-Parts*)

Drawdio (*http://drawdio.com*)

MaKey MaKey (*http://makeymakey.com*)

RoboBrrd (*http://robobrrd.com*)

Project: Make a littleBits Plotter

What Is a Plotter?

A plotter (Figure 5-1) is a machine that uses a pen or other art tool to draw a line image on paper or other physical objects by following a program or responding to real-time input.

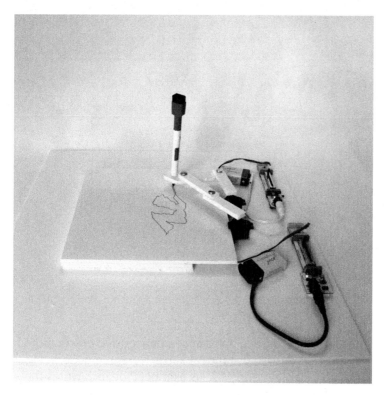

Figure 5-1. *The littleBits plotter in manual Turn mode*

What It Does

Because a pen plotter draws lines instead of printing dots like most modern printers, it can create a more exact copy of the image it is trying to reproduce. It is commonly used for large-scale drawings, such as engineering diagrams and architectural layouts, because it can accommodate long sheets or rolls of paper. And artists are tinkering with plotters because they can use many of the same kinds of pens, brushes, ink, and paint used to create images by hand.

Where It Came From

Plotters generally consist of a pen holder mounted on a set of two rails.

One hobby robot that uses this configuration is the WaterColorBot kit from Evil Mad Scientist (Figure 5-2). The WaterColorBot was designed by 12-year-old Sylvia Todd, star of Sylvia's Super-Awesome Maker Show, a web series of video tutorials, who took her creation to the White House Science Fair in 2013. The WaterColorBot can dip a brush in paint from a child's watercolor set and copy the movements of a human painter in real time, or follow a program created in Arduino. A similar Evil Mad Scientist plotter is the Egg-Bot. The pen in the Egg-Bot can follow the contour of any spherical or ovoid-shaped object that fits in its holder. One fan built a variation, the Mug Marker, which does the same for coffee cups.

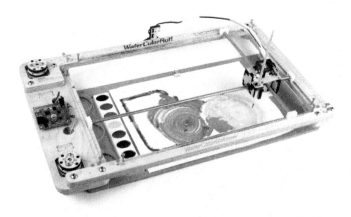

Figure 5-2. *The WaterColorBot can reproduce human-created paintings or images from a program or in real time. Credit: Evil Mad Scientist/WaterColorBot.com.*

Other plotters work like seismographs and move a pen across a paper using one or more arms. The desk toy Plotclock (Figure 5-3), designed at the FabLab in Nuremberg, Germany, holds a dry-erase marker in a double-handed grip. Elbows akimbo, Plotclock writes the time on a teeny whiteboard—then a minute later wipes the slate clean with an eraser and writes the time again. French artist/scientist Patrick Tresset's single robotic arm, dubbed Paul, uses its camera eye to record a subject's face and sketches out a portrait in its human developer's own scribbly style. This style of plotter also includes the Pancake Bot, which we discussed earlier.

Some plotters can be configured to produce "original" art by reinterpreting images through their programming or the individual quirks of the drawing mechanism itself. The wall plotter, also called a vertical plotter, or v-plotter for short, is a pen holder suspended on a string from the corners of an upright drawing surface. Servo motors let out or reel in the string to raise and lower the pen and swing it from side to side. Unlike other kinds of plotters, it doesn't matter how big the paper is—as long as there's enough string, it can keep going. A group called Norwegian Creations built one of the first hanging drawbots using filled soda bottles as weights. Other examples include Der Kritzler from Germany, which writes on windows, and Hektor from Switzerland, which creates lovely robotic graffiti using spray paint.

Figure 5-3. *The Plotclock is an Arduino-controlled plotter arm that writes down the time each minute on a white-board. (Credit: thingiverse.com/joo[Johannes Heberlein])*

How It Works

Like 3D printers, laser cutters, and programmable saws, sewing machines, and other CNC devices, plotters use CAD software to turn photos or graphics into computer code. A program translates the image into points on a grid, and then tells the plotter how to move to get the pen into the right place. Plotters that use Cartesian geometry plot the points on an x-axis (horizontal) and y-axis (vertical). It's just like the geometry you learned in high school, when you drew lines and curves on a piece of graph paper by counting off boxes from the place where two guide lines crossed. Usually a z-axis motor lifts the pen up off the paper and lowers it down again.

Other types of plotters use polar coordinates that calculate the distance from one or more points on the machine. String plotters work this way. The poles are usually at the corners where the string is hung. Plotters with arms may use either, but like string plotters, the programming may be more complex than with straightforward Cartesian coordinates. That's because the image may have to be built up by a pen that makes sweeping curves instead of traveling straight up or down.

Break the Code: Traveling Salesman Problem

The Traveling Salesman Problem is one of the most studied puzzles in mathematics. It goes like this: a traveling salesman has to visit customers in a number of cities and then return back to the home office. What is the shortest path connecting every city on his itinerary that doesn't intersect itself? As the number of cities grows, so does the complexity of the problem. In fact, some mathematicians believe that there is no exact answer, only approximations. Most often, the solution looks like a wiggly line that doubles back on itself over and over, like a maze.

The Traveling Salesman Problem can also be applied to the kind of plotters that draw a continuous line without lifting the pen from the paper. Instead of cities, the wiggly lines connect points on the drawing. By using programming based on the Traveling Salesman Problem, these kinds of plotters can create areas of dark and light shading by making the wiggly lines closer together or farther apart.

Making the Project

The littleBits Plotter is based on the Plotclock, the small drawing robot with a two-fisted grip. The arms are moved by servos, electronic motors that let you control how far the shaft turns and in what direction. Even though there is only one motor for each arm, they each have two degrees of freedom—they can bend at the "shoulder" and at the "elbow." That gives this drawing machine four degrees of freedom in total. The real Plotclock can also tilt the pen up and away from the drawing surface, so it has five degrees of freedom. While the original design is programmed to write and rewrite the time using an Arduino, the low-tech adaptation is controlled by sliding dimmer switches. But this minimalist variation can still produce some interesting movements—both automated and manually controlled.

The littleBits servos you'll be using to build your Plotclock look-alike have two modes, "Turn" and "Swing." In Turn mode (Figure 5-4), the servo can be steered using a dimmer switch that controls the voltage of the circuit—turning to the left when the voltage is low, and to the right when the voltage is high. The sliding dimmer fits this project particularly well because the motion of your hand on the slider matches the motion of the servo back-and-forth. (The littleBits dimmer knob also works.) In Swing mode (Figure 5-5), the servo oscillates back and forth on its own and the dimmer controls the speed of the servo arm's swing.

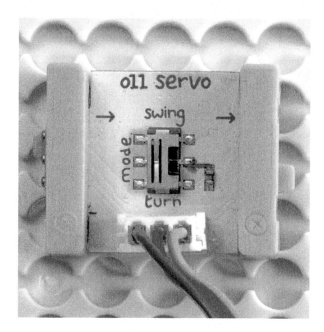

Figure 5-4. *The littleBits servo in Turn mode.*

Figure 5-5. *The littleBits servo in Swing mode.*

Don't forget to document your work!

Project Parameters

- Time Needed: 1–2 hours
- Cost: $145 (about $15 for craft materials and $130 for reusable littleBits modules purchased individually or in a kit)
- Difficulty: Easy
- Safety Issues: A sharp box cutter or art knife is needed to cut the foam core

What You Need to Know

- Skills You Already Have: Measuring, cutting, and gluing
- Skills You Will Learn: Prototyping, building circuits, designing moving parts

Gather Your Materials

- Drawing mechanism and platform
- Sharp pencil for drawing lines and making holes in the foam core board
- Metal or metal-edged ruler
- Foam core board, enough for one 16 inch (40 cm) square and several smaller pieces

- Dry-erase foam core board, enough for one 8-inch (20 cm) square (wide packing tape or slick peel-and-stick shelf paper on regular foam core works, too)
- Three brads (two-prong metal paper fastener)
- Dry-erase markers (fine point, such as Quartet or Board Dude brands)
- Box cutter or X-Acto art knife
- Small adhesive dots or white glue
- Clear (gel-like) mounting tape, 1 inch (25.4 mm) wide

 See "Building Robots with littleBits" on page 45 for more information about using removable adhesives that let you reuse your littleBits modules in other projects.

littleBits Modules

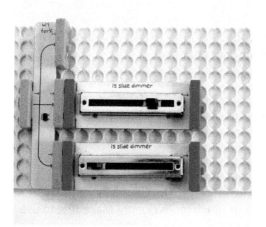

Figure 5-6. *A littleBits Fork and two Slide Dimmers on a Mounting Board*

- 9V Battery and Cable (SKU 660-0006)—you need two, *or* one plus a splitter, such as the Split double wire or triple-branched Fork
- Power Module (littleBits #LB-BIT-p1-POWER-v03)—you need two, *or* one plus a splitter, such as the Split double wire or triple-branched Fork
- Slide Dimmer (SKU 650-0110) *or* Dimmer (SKU 650-0122)—two total
- Servo Motors (# 650-0041)—two preferable, but see Step 7 for a one-servo variation

Optional littleBits Modules

- Mounting Boards two-pack (#660-0005)
- Pulse (littleBits #LB-BIT-i16-PULSE-v03)
- Light Sensor (littleBits #LB-BIT-i13-LIGHTSENSOR-v03)
- Inverter (littleBits #LB-BIT-w10-INVERTER-v03)
- Randomizer (littleBits #650-0129)

Directions
Step 1: List Your Requirements

The goal of the project is to build a drawing machine modeled on the Plotclock that can move a pen around on its own or be controlled by littleBits electronic switches.

Step 2: Plan Your Project

There are lots of ways to use the Plotclock's simple but effective arm design to create a drawing machine. Assembling it on a test platform gives you maximum flexibility to try out different configurations. The platform also gives you a place to anchor all your parts. Keep in mind that the servo motor bodies need to be anchored securely in place, so the only thing that moves when you turn them on are the rotating arms. You'll also need to make sure the foam core arms are connected firmly to the servo arms, so they don't wiggle or come loose as the mechanism starts to move.

 For free files to make your own 3D-printed holders to attach the battery and the servos to the littleBits mounting boards, see "Plotter Linkbox" on page 153.

Step 3: Stop, Review, and Get Feedback

This build is pretty straightforward and forgiving. There should be plenty of foam core if you make a mistake or want to try a different design. Pro tip: Elmer's Dry Erase board is more expensive than a regular board, but cuts much more cleanly. Make sure to use a fresh blade on your box cutter or art knife. Use a metal ruler to guide you, and make several slow, steady swipes until you cut through all the layers, rather than bearing down on the blade.

Note that the measurements given below (in inches and centimeters) are taken from my prototype. A little deviation here or there should not affect it. If you want to make significant changes to the proportions, however, you'll need to prototype your version yourself to be sure that it all fits together.

The only component you may have trouble with is the servo. Be careful when changing the arms to avoid loosening or bending them. Keep track of the little screw that holds the arm on. One handy trick is to place the servo on a dish or paper plate before you disassemble it. That way any small pieces from the servo will stay on the paper plate, and not roll off the table and get lost.

Step 4: Build Your Prototype

1. Measure and cut the following pieces from any kind of foam core board:

   • A base about 16 inches square for the base.

   • Two 8 inch (20 cm) by 6 inch (15 cm) rectangles to support the drawing surface.

   • Two rectangular arm segments 1/2 inch (1 cm) wide by 2 1/2 inches (6 cm) long.

   • One rectangular arm segment 1/2 inch (1 cm) wide by 3 inches (8 cm) long. Snip off the corners at one end to round it off a little.

   • Cut one 8 inch (20 cm) square for the drawing surface, using the dry-erase board. You can also use regular foam core board and cover it with dry-erase peel-and-stick paper or packing tape. Try to keep the surface as smooth as possible, making seams match exactly and avoiding air bubbles.

2. Looking at the arms from the point of view of the machine, the pen will be held in a hand attached to the right arm. It looks like a flag hanging on a pole. The bottom of the segment is a rectangle 1/2 inch (1 cm) across. One side is 2 3/4 inches (7 cm) long. The other side is 2 inches (5 cm) long. The bottom of the "flag" on the "pole" meets the shorter side at a 135 degree angle. The "bottom edge" of the flag is 1 1/4 inches (3 cm) long. The top of the "flag" is parallel to the bottom and the same length. The end of the flag is perpendicular to the top and bottom, about 1 inch (2.5 cm) across.

3. Take the two shorter arm segments. At their upper ends, use a sharp pencil to poke a small hole about 3/8 inch (1 cm) down, centered from side to side. Make the hole just big enough for the metal prongs of the brad to fit through and turn without scraping the sides of the foam core. Do the same at both ends of the longer rectangular arm segment.

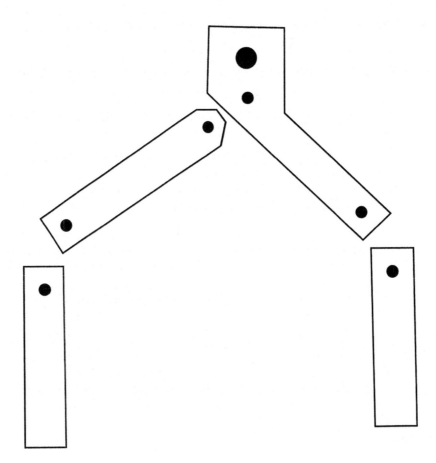

4. On the pen holder arm segment, poke one hole 3/8 inch (1 cm) from the bottom. On the "flag" end, poke another hole 1 inch from the top and centered, about 5/8 inch (1.5 cm) from either side. To make a hole for the pen, make a hole about 3/8 inch (1 cm) down from the top. Use the pencil to widen it until the pencil can almost pass through it.

5. Assemble the left arm by lining up the holes in one short segment and the bottom (nonrounded-off) end of the long segment. Make sure the short segment is underneath! Push the prongs of a brad through the guide holes.

6. Check to see that the segments can pivot freely around the prongs. If necessary, use the pencil to widen the hole a little. When the fit is acceptable, bend the prongs all the way back so they are flat against the back of the foam core. Then shorten the prongs by folding them in half so that the tips meet at the hole. Flatten as much as possible.

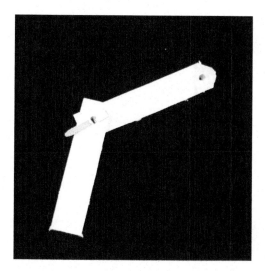

7. Assemble the right arm by lining up the holes in the remaining short segment and the bottom of the flagpole segment, with the short segment on the top (the reverse of the left arm). Fasten them with a brad as before. Finally, attach the two arms by lining up the remaining guide holes with the left arm over the right arm. Be sure the pen holder is underneath the other arm. Again, fasten with a brad the same way.

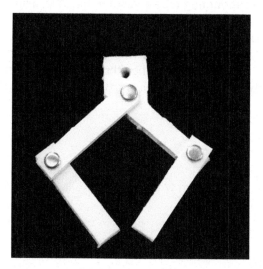

8. Next, build the drawing surface. Glue the two supporting rectangles together, one on top of the other. Then glue the dry-erase square on top—shiny side up—lining up three edges with the bottom supporting pieces so the extra hangs over on the fourth side.

9. Now prepare the servos. The single-prong attachment is the best one for this project. If it's not already on, carefully unscrew and remove the attachment that is there and replace it with the one-pronged servo arm. Don't put the screw back in yet—first you're going to make sure the arms swing across the drawing area. If you've got mounting boards and a 3D-printed holder for your servos, attach them now. If not, take the clear gel-like mounting tape and cut a thin strip, a little narrower than the servo body. Stick it on the bottom of one servo. Attach the servo to the foam core platform about 3 1/2 inches from the edge closest to you, with the business end of the servo facing towards the center of the square, about 16 3/4 inches in from the left edge. Do the same with the other servo, but measuring it in from the right edge. Press the servo bodies down into the tape to attach them as firmly as possible.

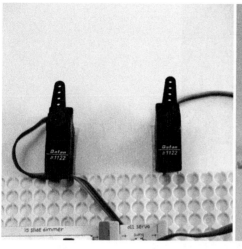 

10. To adjust the swing, complete the circuit by connecting the batteries and power modules (or single battery/power module and a splitter) to the dimmer switches. Place one dimmer on the left and one on the right, lined up horizontally. The battery end of both setups should be pointing the same way (e.g., on the left with the dimmer on the right). Make sure the servos are in Turn mode by checking the little switch on the servo littleBit, and power them up. Push the left slider all the way to the left and the right slider all the way to the right. (If you are using the rotary dimmers instead of the sliders, turn the knobs accordingly.) Notice where the servo arms are pointing; they should be at 3 o'clock and 9 o'clock. Now push the sliders to the other extreme. Make sure the arms are close, but not so close that the foam core arms bump into each other when attached to the servo arms. If necessary, pull the servo arms off and set them at a better angle. When you're satisfied, screw them in securely.

11. Attach narrow strips of the clear mounting tape to the undersides of the foam core arms. Press them down into the servos firmly enough to squeeze some of the tape into the holes on the servo arm. If needed, attach a second piece of gel tape to the underside of the servo arm and press upward into the foam core, sandwiching the servo arm in.

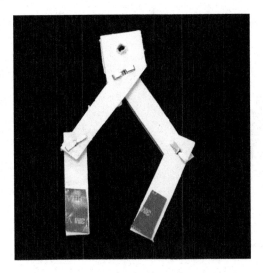

12. If you're using a mounting board, attach it to the platform with a little more tape. Build your littleBits circuit by connecting the battery to the power module to the sliding dimmers and then to the servos.

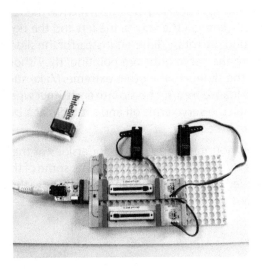

13. Slide the drawing surface towards the servos so the lip is centered up against their fronts. It should fit snugly under the little lip on the servos. You can mark on the platform where you want the drawing surface to sit, but it should stay put fairly well without having to glue it down. That makes it easier to slide it out to wipe it clean. Now, give it a test run!

Step 5: Test Your Design

Take the cap off one of the dry-erase markers and push the point down through the large hole in the pen holder. It should fit snugly and sit with the felt tip just touching the drawing surface. Set the servos to Turn and switch on the power. Try manipulating the sliders to move the pen back and forth, up and down. See if you can get it to make circles, loops, or lines. Can you draw a shape, a picture, or a letter?

To make it draw automatically, turn off the power, set the servos to Swing, and turn the battery back on. Watch what designs they make with the sliders at various settings. If you've got pulse modules, change the servos back to Turn mode, and see how the combination of pulse and slider affects the automatic drawings.

Step 6: Troubleshoot and Refine

The most common problems you are likely to have in running your drawing machine is to keep the foam core pieces connected to the littleBits and vice versa. More gel tape (and pressing the pieces together harder) is usually the answer. If needed, you can try using small nuts and bolts to hold the foam core onto the servo arms.

If your circuit isn't working, make sure all the littleBits pieces are making a good connection with each other. The magnets should be touching. Sometimes the pulse setting will make the arms swing into each other. Try adjusting the relative settings of the pulse modules and the dimmers to try to avoid that, or move the servos farther apart.

Step 7: Adaptations and Extensions

A simpler and cheaper adaptation is to attach the arms to a foam core stand instead of the servo motors and move them manually, like a puppet. Attach them to the stand with brads, and make a knob from foam core or thumbtacks if desired. If you only have one servo, you can try a half-and-half approach by powering one arm and letting the other arm pivot freely on a similarly sized stand.

Other variations include moving the servos closer together or farther apart, or changing the proportions of the arm segments. (This is easier to do if you use the littleBits mounting boards and make 3D-printed battery and servo holders to fit.) Widen the "hand" of the drawing arm to accommodate two or three pens for a multicolor effect. Or instead of a reusable dry-erase board, substitute standard ink pens and paper. Use a large-size pad of sticky notes, or hook up the littleBits DC motor to make a roller that pulls a strip of paper across the drawing surface. If you connect the ends of the strip with tape, you can make a loop that circles around the drawing surface, letting you add multiple lines to the design.

To extend the project further, try adding one or more of the pulse, light sensor, inverter, or randomizer modules to change the way the arms are controlled. littleBits also has Arduino modules that let you add programming (see "Project: Make FiberBot, an E-Textile Arduino Robot" on page 153 for a look at how Arduino works). Or if you're skilled in working with electronic components, build your own permanent circuit with servos and dimmer switches.

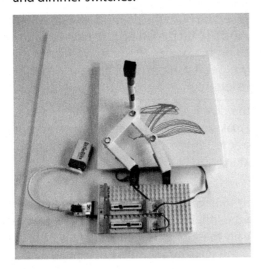

Plotter Linkbox

WaterColorBot (*http://watercolorbot.com*)

Egg-Bot (*http://egg-bot.com*)

Mug Marker (*http://bit.ly/mug-marker*)

Plotclock   (*http://www.thingiverse.com/thing: 248009*)

Patrick Tresset's robot Paul (*http://bit.ly/p-tresset*)

Der Kritzler (*http://bit.ly/der-kritzler*)

Hektor (*http://juerglehni.com/works/hektor*)

PlotterBot website listing many drawing robots (*http://plotterbot.com*)

littleBits modules (*http://littlebits.cc/products*)

Battery holder (*http://www.thingiverse.com/thing: 243530*)

Servo holder (*http://www.thingiverse.com/thing: 243545*)

Project: Make FiberBot, an E-Textile Arduino Robot

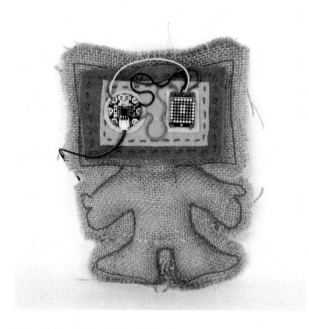

Figure 5-7. *A finished FiberBot.*

What Are E-Textiles?

E-textiles combine sewable electronic components and fiber art.

What It Does

E-textiles can receive input from sensors and switches and react with the physical environment by moving, flashing lights, producing sounds, and so on, using Arduino-compatible hardware and Arduino programming.

Where It Came From

The Arduino microcontroller board was developed in 2002 by software designer Massimo Banzi for his students at the Interaction Design Institute Ivrea in Italy. His goal was to create a tool that would help nontech-savvy students create electronics projects as quickly and easily as possible. There were already a few microcontrollers being used for physical computing—making physical objects respond to commands and interact with their environment. But Banzi wanted something cheaper, more powerful, and compatible with Mac and Linux operating systems as well as Windows. With the help of a team of students and colleagues, Banzi based the Arduino (named after their favorite local bar) on a programming language from MIT called Processing that helped beginners create interesting graphics. He borrowed Processing's easy-to-use integrated development environment (IDE), and adapted it to work with a specially designed microcontroller board that could work with lights, motors, and other physical outputs. To speed up the development process, Banzi's team also made Arduino open source, which means anyone can build their own or come up with their own variation.

Today Arduino is one of the most popular tools for robotics hobbyists and makers of all kinds. One of the things hobbyists like to do with Arduino is use it to add programmable controls (and often sensors and processors to allow them to operate autonomously) to old toys. The small size and low cost of Arduino boards and Arduino-compatible clones make it possible to embed them into RC cars, stuffed dolls, and more. Many Arduino-compatible systems attempt to miniaturize the components even more. Some, like TinyDuino, do it by creating modular systems that let you choose only those features you need and connect them by stacking them together. Others make all-in-one boards but usually use a smaller microcontroller, and limit the number of input/outputs they have, to save space.

That's the route taken by pared-down sewable microcontroller boards. Designers use Arduino to create clothing and decorative accessories that can respond to their environment or your control. They use conductive thread and fabric to build circuits right into the fabric of their design, often in visible and decorative ways. And they incorporate sewable microcontrollers like MIT professor Leah Buechley's line of LilyPad components (developed with and manufactured by SparkFun). While most e-textiles are just for fun, some are also useful. For example, Leah Buechley's Turn Signal Jacket can be worn while bicycling to alert drivers behind you. And design student Kathryn McElroy of the School of Visual Arts created a handbag covered in flashing colored lights that lets you know if you forgot your phone or keys. A small Arduino-based circuit board inside the handbag can read a signal from RFID (radio frequency identification) tags you affix to your personal belongings. Then you can program the microcontroller to create different animated patterns, depending on what item it's detecting (or not). You could set it to flash red if it doesn't get a signal from your eyeglass case, for instance.

How It Works

Arduino boards and their clones come with *pins*, which are really sockets where you can plug input devices (like light sensors or dimmer switches) and output devices (light motors and LEDs). Programming an Arduino board involves telling it what pin to look at for input data, and what pin to send output instructions to. The Arduino language itself is relatively easy to understand. It doesn't take long to figure out how to write simple commands and programs, which are known as *sketches*. There are even drag-and-drop graphical versions of Arduino under development, such as Scratch for Arduino and Ardublocks, which make working with Arduino boards even easier.

Sewable Arduino boards like LilyPad Arduino are very versatile. Because they're smaller, simpler, and less expensive than the standard Arduino boards, they're a popular way for beginners who already know traditional crafts like sewing and crochet to add cool interactive features to their fiber art. They feature large holes surrounded by big metal pads, which make it easier to create electrical connections using conductive thread. The trade-off is that the mini boards can't do as much as a full-sized board. The LilyPad Arduino line of components are very good for making lights blink and shine in programmed patterns or in response to stimuli like motion, sound, or the beating of a heart. But they're limited when it comes to more powerful applications such as making a robot move.

In 2013, Adafruit Industries, a company founded by MIT-trained engineer Limor "Lady Ada" Fried, came out with a line of ultra-tiny microcontroller boards that could handle things like servo motors as well as other interesting kinds of components. Like the LilyPad Arduino, and the sewable Flora and Gemma (and the nonsewable Adafruit Trinket) are smaller and thinner than regular Arduino boards. That makes them ideal for for beginning projects and tighter spaces, including programmable jewelry, 3D-printed toys, and ragdoll hacks.

Arduino Linkbox

Arduino (*http://arduino.cc*)

Leah Buechley's LilyPad Arduino (*http://lilypadardui no.org*)

Kathryn McElroy's Chameleon Bag (*http://bit.ly/ chmln-bag*)

Scratch for Arduino (*http://s4a.cat*)

Ardublock (*http://blog.ardublock.com*)

Adafruit (*http://www.adafruit.com*)

- *Getting Started with Arduino* by Massimo Banzi (2011, Maker Media)

- *Make: Arduino Bots and Gadgets* by Kimmo Karvinen and Tero Karvinen (2011, Maker Media)

- *Make: Basic Arduino Projects* by Don Wilcher (2014, Maker Media)

Making the Project

The FiberBot's Borg-like electronic left eye is really an Adafruit Gemma microcontroller. Its animated right eye is an LED screen, also from Adafruit. You'll learn how to program the Arduino-based Gemma to run an EyeRoll animation, or create one of your own. Strictly speaking, this isn't an e-textile project, because the LED matrix isn't set up to use conductive thread. But you'll get to use your sewing skills to create the basic hand-stitched, stuffed FiberBot body and attach the circuit to it. It's a challenging project for a beginner, but when you're done you'll be able to say you know how to work with entry-level Arduino devices.

 The Adafruit computer programs reprinted here are available free to the public through a BSD open source software license. They ask that any use of the code include the following message: "Adafruit invests time and resources providing this open source code. Please support Adafruit and open-source hardware by purchasing products from Adafruit!"

 Don't forget to document your work!

Project Parameters

- Time Needed: 3–4 hours (less for experienced robot builders; more if you're a newbie)
- Cost: $20–$30
- Difficulty: Moderate
- Safety Issues: See "Project: Make a Souped-Up Solar BEAM Wobblebot" on page 83 for safety issues regarding soldering.

What You Need to Know

- Skills You Already Have: Sewing, downloading and unzipping computer files, soldering (see "Project: Make a Souped-Up Solar BEAM Wobblebot" on page 83)
- Skills You Will Learn: Setting up an Arduino device and working with an Arduino program

Gather Your Materials

Note that the number in parentheses indicates the Adafruit product number.

Figure 5-8. *The Adafruit Mini LED Matrix (top left), the Matrix backpack (bottom left), and the Gemma microcontroller (bottom right)*

- Adafruit Gemma microcontroller (#1222)
- Adafruit Mini (0.8 inch) 8×8 LED Matrix with backpack (#872 or other colors in same size)
- Adafruit 3 x AAA battery holder with on/off switch and 2-pin JST connector cable (#727)
- Mini USB cable (#260—a cell phone cable that fits will work, too)
- Computer (preferably Windows or Mac) with Internet
- Wire (preferably solid; 22 AWG or thinner hookup wire, or 30 AWG wire wrap)
- Wire cutter and wire stripping tools
- Soldering iron and solder

The Fabric Body

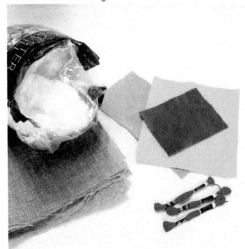

- Paper and pen
- Burlap (used burlap bags are good), enough to make two 8 × 10 inch (20 × 25 cm) pieces
- Felt (three or four colors)
- Straight pins
- Embroidery needle and/or large sewing needle, size 3 or 5—must fit through the holes on the Gemma
- Scissors
- Embroidery yarn
- Embroidery needle (thick, with an eye large enough for a full strand of embroidery yarn)
- Fiber fill stuffing for the body (can also use fabric scraps or dryer lint)
- Peel-and-stick Velcro tape or tabs

A Note on Wire and Batteries

The numbers you see listed for different kinds of wire refer to the thickness, or gauge (sometimes listed as AWG, for American Wire Gauge). The higher the number, the thinner the wire. The holes on some of the components for this project are rather small, so 22 AWG is the thickest you will want to use. That's the gauge of standard hookup wire, an all-purpose wire you can buy insulated in several colors.

Also look for solid wire as opposed to stranded. For soldering components, solid wire is easier to work with. Stranded wire is made up of many ultra-thin filaments that can fray and tear, making it harder to poke the wire through small holes. As a shortcut, you can buy a box of precut, prestripped breadboard jumper wires and have a supply of short wires in multiple colors, already *stripped* (to expose the wire inside the plastic insulation) and ready to go.

When it comes to batteries, there are a few different choices that will work, but for beginners the best choice is the recommended three AAA batteries. The battery holder from Adafruit has a JST plug already attached, as well as an on/off switch, saving you a few more parts to find and solder on. Coin cell batteries, recommended for some other projects using the Gemma, take up less space but don't deliver enough current to power the LED matrix.

You may see that Adafruit recommends a lithium-ion polymer battery for similar projects. The LiPo battery costs just a little more than a AAA battery holder and batteries, but is much smaller, lighter, and rechargeable. However, LiPo batteries also have some safety concerns—their covering is soft, and they can catch fire if bent, crushed, punctured, or short-circuited. They also need to be recharged carefully, preferably with a special charger like the USB model Adafruit sells. That's why we don't recommend using a LiPo battery for beginners.

Directions
Step 1: List Your Requirements

There are two goals to work on in creating this project: making a soft "robot body" from fabric, and integrating an Arduino board and some electronics to kick it up a notch. Getting those two goals to work together well is the challenge. This version is just one solution—once you understand a little of the technology involved, feel free to give this project your own stamp.

Step 2: Plan Your Project

Use the pattern in this book or design your own. Just make sure the "face" of the FiberBot fits on the body you choose. All the electronics will get sewn onto the rectangle-shaped face, which gets attached to the FiberBot head with Velcro, so you can take it off to program it or to use on an alternate body. The battery pack plugs into the Gemma with a cable long enough to hang on the FiberBot's back with more Velcro. If you're designing your own body, make your own pattern using a heavy marker that will be visible through the burlap.

Step 3: Stop, Review, and Get Feedback

Setting up the Gemma for the first time and initializing the LED matrix can be confusing. If you need help, check out the Adafruit Learning System online tutorials. There's much more information there than you will need to know to get this project up and running, but that's the first place to turn. Adafruit also has a Customer Support Forum that you can join to get answers to your questions from the company's tech experts and experienced customers. It helps to look through the tutorials first in order to frame your question. You should get a response within a few hours.

Adafruit Tutorials and Support

- Gemma setup (*http://learn.adafruit.com/ introducing-gemma/introduction*)

- Drivers for Windows computers (*http:// learn.adafruit.com/usbtinyisp/drivers*)

- Installing Arduino (*http:// learn.adafruit.com/introducing-gemma/ setting-up-with-arduino-ide*)

- Setting up an Arduino Library (*http:// learn.adafruit.com/adafruit-all-about- arduino-libraries-install-use*)

- LED Matrix animation how-to (*http:// learn.adafruit.com/trinket-slash-gemma- space-invader-pendant/source-code*)

- LED Matrix animated robot face code (*http://learn.adafruit.com/3d-printed-led- animation-bmo/overview*)

- Adafruit Customer Support (*http:// forums.adafruit.com/index.php*)

Step 4: Build Your Prototype
First, assemble the electronic circuit.

1. The LED Matrix comes with a "backpack" circuit board that lets you control its 64 pixels with some simple Arduino programming (which you'll learn more about below). The two must be soldered together. Put the LED Matrix face down on your work surface so the 16 thin wire leads are sticking up. To hold it steady, you can rest it between the jaws of a clamp or pliers (they don't need to be tight). Then take the circuit board—with the black rectangular chip facing up, away from the LED Matrix—and slide it down onto the LED Matrix by carefully matching up the leads with the 16 holes along the circuit board's sides. It doesn't matter which side is left or right. Next, heat up the soldering iron, tin the tip, and touch it to the first lead for a second or two.

   Then push the solder underneath the tip just until it flows around the wire into a volcano shape. Let it cool for a few seconds, then continue down the rows on both sides the same way, being careful not to let the solder of one lead run into the next. If it does, see the discussion on desoldering in "Soldering Cheat Sheet" on page 88 to find out how to clean it up and prevent bridging between two points. When you are done soldering all the leads, use wire cutters to trim each one back to just above the little cone of solder.

2. Next you'll use the hookup wire to connect the LED Matrix and backpack to the Gemma. Cut four 4 inch (10 cm) lengths, preferably different colors. Strip the insulation off about a 1/4 inch (1 cm) from each end. Use one arm of your helping hand tool to hold the LED Matrix flat, and the other to hold the first wire sticking up through the first hole. Solder as before. Do the same with the rest of the wires. When done, trim all the leads.

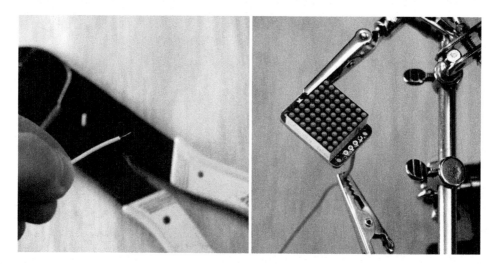

3. Now it's time to connect each wire from the LED Matrix to the Gemma. Looking at the LED Matrix with the lights facing up, notice that the holes for the wires are marked "+," "," "D," and "C." The Gemma has six soldering pads, three on each side. You will be using four and leaving two empty. With the Gemma face up, its USB connector at the top, this is how the wires will be attached.

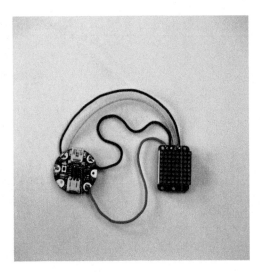

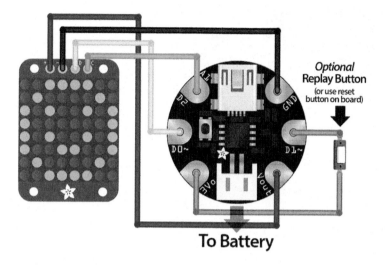

To Battery

Figure 5-9. *Credit: Adafruit and Fritzing.*

- The + wire (red in the photo) will be connected to the bottom pad on the right (marked Vout).
- The – wire (black in the photo) will be connected to the top pad on the right (marked GND).
- The D wire (yellow in the photo) will be connected to the middle pad on the left (marked D0).
- The C wire (green in the photo) will be connected to the top pad on the right (marked D2).

Soldering the wires from the LED matrix onto the Gemma is a bit tricky, since the holes are sized for sewing, not soldering, so it helps to pre-tin them. Use the third hand tool to hold the Gemma in place, and cover the pads listed above (and only those pads) with a thin layer of solder. Then take the first wire and poke it through the correct hole from below, double-checking that you have the right one. Bend the wire over to hook it onto the first pad. Use the tip of the soldering iron to push the wire down into the blob of solder on the appropriate pad, adding a little more from the top until the wire is covered. Do the same with the other three wires.

4. At this point, you may want to go to "Step 5: Test Your Design" to load the Arduino software onto your computer, plug in the Gemma, and make sure it and the LED Matrix are working properly. Otherwise, you can finish building the body of your FiberBot and work on the programming later.

Next, make the FiberBot body and attach the Gemma-LED Matrix circuit to it.

1. FiberBot's raggedy burlap body was inspired by the little androids in the movie *9* by Tim Burton protege Shane Acker. Burlap is inexpensive and easy to cut and sew, but you'll need to take care to prevent fraying. Print out the FiberBot Pattern (*http:// bit.ly/fiberbot-template*) on regular-sized printer paper (or enlarge and copy the pattern below) and place it over your pattern. The height of the bot from head to foot should be 8 inches (20 cm). Trace the outline of the bot onto the burlap with a soft pencil. Place it on top of the other piece of burlap, matching the edges, but don't cut it out yet!

2. You will be sewing around the outline of the FiberBot body using the backstitch, starting on one side of the opening at the top of the head marked with a dotted line on the pattern. To do a backstitch, thread the largest needle you have with a piece of embroidery thread about 3 feet (1 m) long. (Do not separate the thread into strands.) Bring the needle through from the back to the front about 1/8 inch (3 mm) ahead of where you want to start. Go back to the starting point and push the needle through from front to back. Take one more stitch the same way to anchor the thread. Then come around once more, but this time push the needle up from back to front about 1/8 inch (3 mm) ahead of the end of the previous stitch. Then go back to the end of the previous stitch and push the needle through from front to back. Continue on with the backstitch around the entire body, stopping on the other side of the opening. Leave any excess thread hanging for now.

3. Take the felt and cut out two large rectangles, one medium rectangle, one small rectangle, and one circle, following the pattern. Center the medium rectangle on one of the large rectangles and sew them together using a running (in and out) stitch. Make a knot at the beginning and the end of your sewing to keep the thread from coming out. Trim any excess thread at the end.

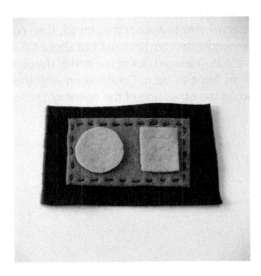

4. The Gemma and the LED Matrix go side by side on top of the medium rectangle, with the Gemma on the left. If the wires connecting them are stiff enough, curl them (gently!) into decorative waves to take up the slack. Be careful not to bend them too sharply, especially at the soldered ends, or they will break. Position the small felt rectangle under the LED Matrix. Using the large holes on the corners of

the LED backpack, sew the device onto the felt, through all the layers, using three stitches in each hole. Knot the thread at the beginning and end to keep it in place. There are two holes available on the Gemma for sewing. Position the circular felt pad underneath the Gemma and sew it on through those two holes the same way as the LED Matrix.

5. The second large felt rectangle is to hide all the stitching on the back of the first large felt rectangle. Attach four small peel-and-stick Velcro tabs onto the corners of the second felt piece, leaving room around the edges for stitching. Sew the felt pieces together, with the Velcro tabs on the outside. Attach the corresponding pieces of Velcro to the FiberBot's burlap head, pressing to make a good seal. Connect the felt face to the FiberBot.

6. Plug the battery pack into the Gemma using the JST connector. Stick a piece of Velcro on the battery case opposite the side with the on/off switch. Stick the corresponding piece to the FiberBot's back, pressing firmly. Now remove the face and the battery pack and set aside. If you have trouble unplugging the JST plug, use a pair of pliers to gently pull it out of the socket.

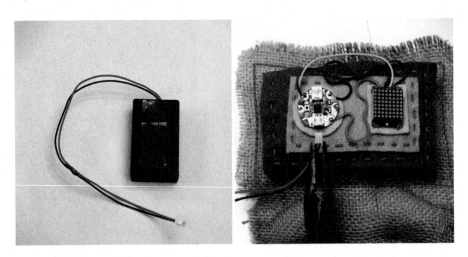

7. Insert the stuffing into the FiberBot through the opening at the top of the head. Use the eraser end of a pencil, a chopstick, or the handle of a wood spoon to get the stuffing into any corners you can't reach with your fingers. Try to fill it loosely —just enough to give it a little three-dimensionality—since the burlap may fray and tear if stretched. When you're satisfied, use the remaining thread to close up the hole with a backstitch.

8. Carefully trim around the outside of your stitching, leaving at least 1/2 inch (1 cm) of excess. You want a little fraying for a rough effect, but don't trim too close to the stitching or the burlap will fall apart. Just take a small snip in tight places like around the arms and the legs, and don't cut between the legs at all. If you do have fraying, you may be able to catch it with another row of stitching. Luckily, this style lends itself to a rough finish. Now reattach all the electronics, and your FiberBot is finished!

Step 5: Test Your Design

When your Gemma and LED backpack are all wired up, it's time to try them out with the Arduino IDE and Adafruit programming. There are several steps involved, and you will have to download the necessary software from the Adafruit website onto your computer. Then you will upload the software from your computer to the Gemma via the USB cable. Here are the steps:

1. If you have a Windows computer, you first need to download a driver called USB-tinyISP. (A driver is a small piece of software that tells the computer how to communicate with the hardware you are plugging in, in this case the Gemma.)

   To install the driver:

   - Go to the "Drivers" section of the Adafruit website (*http://bit.ly/ada-drivers*) for directions for Windows 7, 8, and XP.

   - Plug your Gemma into a USB port on your computer using the Mini USB cable. The onboard green LED should light up, and the onboard red LED should start flashing.

   - An installation wizard pop-up box should open up on your computer screen; tell it not to search the web for a driver, and it should install the driver on your hard drive.

     If the driver doesn't work, try one of the other versions you can find on the Adafruit page. My computer did not automatically open the installation wizard, but I was able to find the device and instruct the computer to manually install the driver folder. You may have to tell your computer to ignore a warning that the file isn't safe; it is.

2. Next, you need to install the version of the Arduino IDE that works with Adafruit's Gemma, Flora, and Trinket Arduino boards. If you already have a standard version of Arduino, there are directions to modify it yourself—but it's easier to just install Adafruit's customized version. Here's how:

- Go to the Adafruit website (*http://bit.ly/arduino-setup*) for instructions and links.
- When you have downloaded the appropriate ZIP folder, move it where desired on your hard drive, and then extract all the files inside.

3. Find the Arduino IDE (it's marked with a little teal Arduino symbol) in the directory where you stored the Arduino files. Click on it and a little pop-up window will open with an Arduino editing screen. Before you run your first sketch, or program, you need to adjust some settings. Click on Tools on the menu bar at the top of the Arduino window and then click on these:

- Under Board, select Adafruit Gemma 8MHz
- Under Programmer, select USBtinyISP
- Under Serial Port, select the number of the USB port you have plugged the Gemma into. (You may need to try a few to find the right one.)

4. To test everything out, you will open and run a little a sketch, called Blink. It's the "Hello World" for Arduino. Here's how:

- Copy and paste the Blink sketch (below and at on the Arduino website (*http://bit.ly/ arduino-setup*), scroll down to the bottom of the page) into the white input screen in the Arduino pop-up window.
- Take the Gemma and press the reset button that sits between the red and green LEDs on the board. The red LED should shine dimly and then start flashing rapidly. That means the bootloader is ready to let you boot up a program.
- Click the Upload icon on the Arduino screen (the arrow that points to the right) and watch what happens. The code should start compiling (meaning Arduino is reading the program and translating it into digital code that the hardware can understand). If all goes well, the red LED on the Gemma should start blinking on and off in one second intervals (more slowly than it was flashing before).

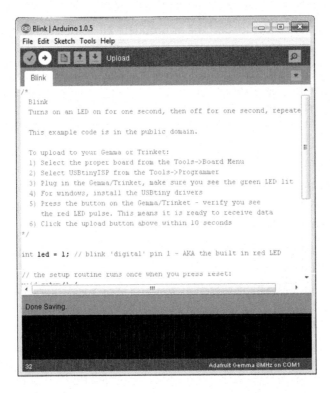

- You can play around with the code for the Blink sketch to see how it works. Look for the lines that say delay (1000); that command tells the Arduino device to turn the LED on or off for 1,000 thousandths of a second (or one second). If you change the numbers to 2,000, it will blink twice as slow. Change them to 500, and it will blink twice as fast. Don't forget to push the reset button on Gemma before you click Upload on the Arduino screen every time.

5. Below the white input screen in the Arduino pop-up window, there is a black status area that tells you if your sketch is running correctly or has errors. If you see red text in that box, that means something's wrong. The most common error message you are likely to get is avrdude: Error: Could not find USBtiny device (0x1781/0xc9f). That happens when you forget to press the reset button on the Gemma before you click Upload. Just try it again and the error message should go away. For other error messages, read through the directions again, and check the page "Setting Up with Arduino IDE" page (*http://bit.ly/arduino-setup*) for any updates. You can also head over to the Adafruit Customer Support Forum, where a live person will answer your questions relatively quickly. However, even with an error message the sketch may run correctly. For purposes of getting your Gemma and LED Matrix up and running, that may be good enough to move to the next step.

6. Once your Gemma is running the Blink sketch successfully, you need to get the LED Matrix involved. For that, you need to download a *library*, or bunch of related programs, called TinyWireM. Here's how:

   • Close any open Arduino windows.

   • Go to the Adafruit website (*http://bit.ly/gemma-sc*) to find the link to the TinyWireM library and download it.

- Unzip the library file. It's important to move it to the correct folder. In Windows, the Arduino IDE will automatically create a *libraries* folder in your *Documents* directory, under *My Documents*. If you need help finding the right folder, check out the Adafruit website (*http://bit.ly/install-lib*) for help.

7. Now you're ready to run some animations on the LED Matrix. The sketches to run the LED Matrix are split into two pages. When you open the pages in the Arduino IDE, they will go in two separate tabs in the editing screen, like the tabs on an Internet browser. The first tab is the source code (Example 5-1). This can stay the same for every animation that you write. To load it, open the Arduino IDE and start a new sketch. Copy the source code below (or find it on the Gemma section of the Adafruit site (*http://bit.ly/gemma-sc*)) and paste it into the Arduino IDE editing screen. Click Save As in the File menu and give the source code a name (like "LED-MatrixSourceCode"), so you have a clean copy you can reuse to create new sketches.

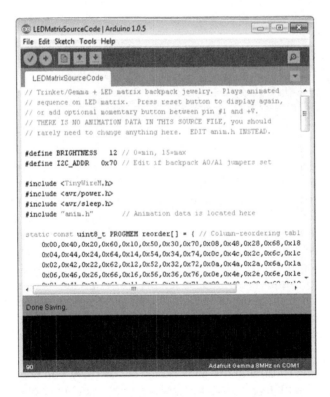

Example 5-1. *Source code for Gemma—LED Matrix*

```
// Trinket/Gemma + LED matrix backpack.  Plays animated
// sequence on LED matrix.  Press reset button to display again,
```

```
// or add optional momentary button between pin #1 and +V.
// THERE IS NO ANIMATION DATA IN THIS SOURCE FILE, you should
// rarely need to change anything here.   EDIT anim.h INSTEAD.

#define BRIGHTNESS   12 // 0=min, 15=max
#define I2C_ADDR   0x70 // Edit if backpack A0/A1 jumpers set

#include <TinyWireM.h>
#include <avr/power.h>
#include <avr/sleep.h>
#include "anim.h"      // Animation data is located here

static const uint8_t PROGMEM reorder[] = { // Column-reordering table
    0x00,0x40,0x20,0x60,0x10,0x50,0x30,0x70,0x08,0x48,0x28,0x68,0x18,0x58,0x38,0x78,
    0x04,0x44,0x24,0x64,0x14,0x54,0x34,0x74,0x0c,0x4c,0x2c,0x6c,0x1c,0x5c,0x3c,0x7c,
    0x02,0x42,0x22,0x62,0x12,0x52,0x32,0x72,0x0a,0x4a,0x2a,0x6a,0x1a,0x5a,0x3a,0x7a,
    0x06,0x46,0x26,0x66,0x16,0x56,0x36,0x76,0x0e,0x4e,0x2e,0x6e,0x1e,0x5e,0x3e,0x7e,
    0x01,0x41,0x21,0x61,0x11,0x51,0x31,0x71,0x09,0x49,0x29,0x69,0x19,0x59,0x39,0x79,
    0x05,0x45,0x25,0x65,0x15,0x55,0x35,0x75,0x0d,0x4d,0x2d,0x6d,0x1d,0x5d,0x3d,0x7d,
    0x03,0x43,0x23,0x63,0x13,0x53,0x33,0x73,0x0b,0x4b,0x2b,0x6b,0x1b,0x5b,0x3b,0x7b,
    0x07,0x47,0x27,0x67,0x17,0x57,0x37,0x77,0x0f,0x4f,0x2f,0x6f,0x1f,0x5f,0x3f,0x7f,
    0x80,0xc0,0xa0,0xe0,0x90,0xd0,0xb0,0xf0,0x88,0xc8,0xa8,0xe8,0x98,0xd8,0xb8,0xf8,
    0x84,0xc4,0xa4,0xe4,0x94,0xd4,0xb4,0xf4,0x8c,0xcc,0xac,0xec,0x9c,0xdc,0xbc,0xfc,
    0x82,0xc2,0xa2,0xe2,0x92,0xd2,0xb2,0xf2,0x8a,0xca,0xaa,0xea,0x9a,0xda,0xba,0xfa,
    0x86,0xc6,0xa6,0xe6,0x96,0xd6,0xb6,0xf6,0x8e,0xce,0xae,0xee,0x9e,0xde,0xbe,0xfe,
    0x81,0xc1,0xa1,0xe1,0x91,0xd1,0xb1,0xf1,0x89,0xc9,0xa9,0xe9,0x99,0xd9,0xb9,0xf9,
    0x85,0xc5,0xa5,0xe5,0x95,0xd5,0xb5,0xf5,0x8d,0xcd,0xad,0xed,0x9d,0xdd,0xbd,0xfd,
    0x83,0xc3,0xa3,0xe3,0x93,0xd3,0xb3,0xf3,0x8b,0xcb,0xab,0xeb,0x9b,0xdb,0xbb,0xfb,
    0x87,0xc7,0xa7,0xe7,0x97,0xd7,0xb7,0xf7,0x8f,0xcf,0xaf,0xef,0x9f,0xdf,0xbf,0xff };

void ledCmd(uint8_t x) { // Issue command to LED backback driver
  TinyWireM.beginTransmission(I2C_ADDR);
  TinyWireM.write(x);
  TinyWireM.endTransmission();
}

void clear(void) { // Clear display buffer
  TinyWireM.beginTransmission(I2C_ADDR);
  for(uint8_t i=0; i<17; i++) TinyWireM.write(0);
  TinyWireM.endTransmission();
}

void setup() {
  power_timer1_disable();     // Disable unused peripherals
  power_adc_disable();        // to save power
  PCMSK |= _BV(PCINT1);       // Set change mask for pin 1
  TinyWireM.begin();          // I2C init
  clear();                    // Blank display
  ledCmd(0x21);               // Turn on oscillator
  ledCmd(0xE0 | BRIGHTNESS);  // Set brightness
  ledCmd(0x81);               // Display on, no blink
}

uint8_t rep = REPS;
```

```
void loop() {

  for(int i=0; i<sizeof(anim); i) { // For each frame...
    TinyWireM.beginTransmission(I2C_ADDR);
    TinyWireM.write(0);               // Start address
    for(uint8_t j=0; j<8; j++) {      // 8 rows...
      TinyWireM.write(pgm_read_byte(&reorder[pgm_read_byte(&anim[i++])]));
      TinyWireM.write(0);
    }
    TinyWireM.endTransmission();
    delay(pgm_read_byte(&anim[i++]) * 10);
  }

  if(!--rep) {                // If last cycle...
    ledCmd(0x20);             // LED matrix in standby mode
    GIMSK = _BV(PCIE);        // Enable pin change interrupt
    power_all_disable();      // All peripherals off
    set_sleep_mode(SLEEP_MODE_PWR_DOWN);
    sleep_enable();
    sei();                    // Keep interrupts disabled
    sleep_mode();             // Power down CPU (pin 1 will wake)
    // Execution resumes here on wake.
    GIMSK = 0;                // Disable pin change interrupt
    rep   = REPS;             // Reset animation counter
    power_timer0_enable(); // Re-enable timer
    power_usi_enable();       // Re-enable USI
    TinyWireM.begin();        // Re-init I2C
    clear();                  // Blank display
    ledCmd(0x21);             // Re-enable matrix
  }
}

ISR(PCINT0_vect) {} // Button tap
```

8. The second program you need to load into the Arduino IDE is the animation data itself. The EyeRoll animation below makes it look like your FiberBot is peering around with its one good eye. To create a new tab, look for a little arrow below the menu bar on the right side of the input screen. Click to open the drop-down menu and select New Tab. Name the new page *anim.h*. Then copy the EyeRoll code below (Example 5-2) and paste it into the new page. Click Save As in the File menu and give the sketch a name (like "MyEyeRollAnimation"). If you don't already have a folder for your sketches, create one in the directory with the rest of your Arduino code and save this sketch there. Arduino will create a folder that contains the source code and the animation sketch.

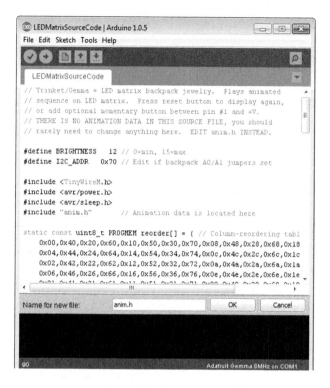

Example 5-2. *EyeRoll sketch*

```
// EyeRoll Animation data for Gemma + LED matrix backpack.
#define REPS 3 // Number of times to repeat the animation loop (1-255)
const uint8_t PROGMEM anim[] = \{

  B00000000, // frame 1
  B00000000,
  B00000000,
  B00000000,
  B00000000,
  B00000000,
  B11111111,
  B01111110,
  10, // 0.1 second delay

  B00000000, // frame 2
  B00000000,
  B00000000,
  B00000000,
  B00000000,
  B11111111,
  B10000001,
```

```
B01111110,
10, // 0.1 second delay

B00000000, // frame 3
B00000000,
B00000000,
B00000000,
B11111111,
B10000001,
B10000001,
B01111110,
10, // 0.1 second delay

B00000000, // frame 4
B00000000,
B00000000,
B11111111,
B10011001,
B10000001,
B10000001,
B01111110,
10, // 0.1 second delay

B00000000, // frame 5
B00000000,
B11111111,
B10011001,
B10011001,
B10000001,
B10000001,
B01111110,
10, // 0.1 second delay

B00000000, // frame 6
B11111111,
B10011001,
B10011001,
B10011001,
B10000001,
B10000001,
B01111110,
10, // 0.1 second delay

B01111110,   // frame 7
B10000001,
B10011001,
B10011001,
B10011001,
B10000001,
B10000001,
B01111110,
300, // 3 second delay

B01111110,   // frame 8
```

```
  B10000001,
  B10001101,
  B10001101,
  B10001101,
  B10000001,
  B10000001,
  B01111110,
  25, // .25 second delay

  B01111110,   // frame 9
  B10000001,
  B10000111,
  B10000111,
  B10000111,
  B10000001,
  B10000001,
  B01111110,
  500, // 5 second delay

  B01111110,   // frame 10
  B10001101,
  B10001101,
  B10001101,
  B10000001,
  B10000001,
  B10000001,
  B01111110,
  10, // 0.1 second delay

  B01111110,   // frame 11
  B10011001,
  B10011001,
  B10011001,
  B10000001,
  B10000001,
  B10000001,
  B01111110,
  10, // 0.1 second delay

  B01111110,   // frame 12
  B10000001,
  B11100001,
  B11100001,
  B11100001,
  B10000001,
  B10000001,
  B01111110,
500, // 5 second delay

  B01111110,   // frame 13
  B10000001,
  B10000001,
  B10110001,
  B10110001,
```

```
  B10110001,
  B10000001,
  B01111110,
  10, // 0.1 second delay

  B01111110,    // frame 14
  B10000001,
  B10000001,
  B10000001,
  B10011001,
  B10011001,
  B10011001,
  B01111110,
  10, // 0.1 second delay

  B01111110,    // frame 15
  B10000001,
  B10000001,
  B10001101,
  B10001101,
  B10001101,
  B10000001,
  B01111110,
  10, // 0.1 second delay

  B01111110,    // frame 16
  B10000111,
  B10000111,
  B10000111,
  B10000001,
  B10000001,
  B10000001,
  B01111110,
  100, // 1 second delay

  B01111110,    // frame 17
  B10001101,
  B10001101,
  B10001101,
  B10000001,
  B10000001,
  B10000001,
  B01111110,
  10, // 0.1 second delay

  B01111110,    // frame 18
  B10000001,
  B10011001,
  B10011001,
  B10011001,
  B10000001,
  B10000001,
  B01111110,
  300, // 3 second delay
```

```
B00000000, // frame 19
B11111111,
B10011001,
B10011001,
B10011001,
B10000001,
B10000001,
B01111110,
10, // 0.1 second delay

B00000000, // frame 20
B00000000,
B11111111,
B10011001,
B10011001,
B10000001,
B10000001,
B01111110,
10, // 0.1 second delay

B00000000, // frame 21
B00000000,
B00000000,
B11111111,
B10011001,
B10000001,
B10000001,
B01111110,
10, // 0.1 second delay

B00000000, // frame 22
B00000000,
B00000000,
B00000000,
B11111111,
B10000001,
B10000001,
B01111110,
10, // 0.1 second delay

B00000000, // frame 23
B00000000,
B00000000,
B00000000,
B00000000,
B11111111,
B10000001,
B01111110,
10, // 0.1 second delay

//rapid blink

B00000000, // frame 1
```

```
B00000000,
B00000000,
B00000000,
B00000000,
B00000000,
B11111111,
B01111110,
10, // 0.1 second delay

B00000000, // frame 2
B00000000,
B00000000,
B00000000,
B00000000,
B11111111,
B10000001,
B01111110,
10, // 0.1 second delay

B00000000, // frame 3
B00000000,
B00000000,
B00000000,
B11111111,
B10000001,
B10000001,
B01111110,
10, // 0.1 second delay

B00000000, // frame 4
B00000000,
B00000000,
B11111111,
B10011001,
B10000001,
B10000001,
B01111110,
10, // 0.1 second delay

B00000000, // frame 5
B00000000,
B11111111,
B10011001,
B10011001,
B10000001,
B10000001,
B01111110,
10, // 0.1 second delay

B00000000, // frame 6
B11111111,
B10011001,
B10011001,
B10011001,
```

```
  B10000001,
  B10000001,
  B01111110,
  10, // 0.1 second delay

  B01111110,    // frame 7
  B10000001,
  B10011001,
  B10011001,
  B10011001,
  B10000001,
  B10000001,
  B01111110,
  100, // 1 second delay

  B01111110,    // frame 7a
  B10000001,
  B10111101,
  B10111101,
  B10111101,
  B10000001,
  B10000001,
  B01111110,
  20, // 0.2 second delay

  B01111110,    // frame 7b
  B10111101,
  B11111111,
  B11100111,
  B11111111,
  B10111101,
  B10000001,
  B01111110,
  20, // 0.2 second delay

  B01111110,    // frame 7a
  B10000001,
  B10111101,
  B10111101,
  B10111101,
  B10000001,
  B10000001,
  B01111110,
  20, // 0.2 second delay

  B01111110,    // frame 7b
  B10111101,
  B11111111,
  B11100111,
  B11111111,
  B10111101,
  B10000001,
  B01111110,
  20, // 0.2 second delay
```

```
      B01111110,    // frame 7a
      B10000001,
      B10111101,
      B10111101,
      B10111101,
      B10000001,
      B10000001,
      B01111110,
      20, // 0.2 second delay

      B01111110,    // frame 7b
      B10111101,
      B11111111,
      B11100111,
      B11111111,
      B10111101,
      B10000001,
      B01111110,
      20, // 0.2 second delay

      B01111110,    // frame 7a
      B10000001,
      B10111101,
      B10111101,
      B10111101,
      B10000001,
      B10000001,
      B01111110,
      20, // 0.2 second delay

      B01111110,    // frame 7
      B10000001,
      B10011001,
      B10011001,
      B10011001,
      B10000001,
      B10000001,
      B01111110,
      100, // 1 second delay

      B00000000, // frame 5
      B00000000,
      B11111111,
      B10011001,
      B10011001,
      B10000001,
      B10000001,
      B01111110,
      10, // 0.1 second delay

      B00000000, // frame 4
      B00000000,
      B00000000,
```

```
B11111111,
B10011001,
B10000001,
B10000001,
B01111110,
10, // 0.1 second delay

B00000000, // frame 3
B00000000,
B00000000,
B00000000,
B11111111,
B10000001,
B10000001,
B01111110,
10, // 0.1 second delay

B00000000, // frame 2
B00000000,
B00000000,
B00000000,
B00000000,
B11111111,
B10000001,
B01111110,
10, // 0.1 second delay

B00000000, // frame 1
B00000000,
B00000000,
B00000000,
B00000000,
B00000000,
B11111111,
B01111110,
10, // 0.1 second delay

B00000000, // frame 2
B00000000,
B00000000,
B00000000,
B00000000,
B11111111,
B10000001,
B01111110,
10, // 0.1 second delay

B00000000, // frame 3
B00000000,
B00000000,
B00000000,
B11111111,
B10000001,
B10000001,
```

```
  B01111110,
  10, // 0.1 second delay

  B00000000, // frame 4
  B00000000,
  B00000000,
  B11111111,
  B10011001,
  B10000001,
  B10000001,
  B01111110,
  10, // 0.1 second delay

  B00000000, // frame 5
  B00000000,
  B11111111,
  B10011001,
  B10011001,
  B10000001,
  B10000001,
  B01111110,
  10, // 0.1 second delay

  B00000000, // frame 6
  B11111111,
  B10011001,
  B10011001,
  B10011001,
  B10000001,
  B10000001,
  B01111110,
  10, // 0.1 second delay

  B01111110,    // frame 7
  B10000001,
  B10011001,
  B10011001,
  B10011001,
  B10000001,
  B10000001,
  B01111110,
  10, // 0.1 second delay

  B00000000, // frame 5
  B00000000,
  B11111111,
  B10011001,
  B10011001,
  B10000001,
  B10000001,
  B01111110,
  10, // 0.1 second delay

  B00000000, // frame 4
```

```
  B00000000,
  B00000000,
  B11111111,
  B10011001,
  B10000001,
  B10000001,
  B01111110,
  10, // 0.1 second delay

  B00000000, // frame 3
  B00000000,
  B00000000,
  B00000000,
  B11111111,
  B10000001,
  B10000001,
  B01111110,
  10, // 0.1 second delay

  B00000000, // frame 2
  B00000000,
  B00000000,
  B00000000,
  B00000000,
  B11111111,
  B10000001,
  B01111110,
  10, // 0.1 second delay

  //end of sequence

  B00000000, //frame 23
  B00000000,
  B00000000,
  B00000000,
  B00000000,
  B00000000,
  B11111111,
  B01111110,
  300, //3 second delay

};
```

9. Now to test the program. If the EyeRoll sketch isn't already open, go into the folder where the two files you just created were saved. Double-click on the source code file (with the *.ino* suffix) to open both. Press the reset button on the Gemma, wait for the red LED to flash, then click Upload. After a few seconds, the EyeRoll animation should start to play on the LED Matrix. Does it work? Congratulations, you're programming in Arduino! (If not, check Step 6 for troubleshooting tips.)

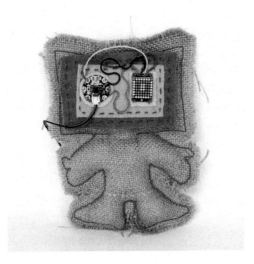

10. You now know how to set up a Gemma and LED Matrix. When you're done, disconnect your FiberBot from the computer and run it off the batteries. If you haven't
already, insert three AAA batteries into the pack and plug it in. To run the animation, turn on the switch on the battery pack. To restart it when it's done, just press
the reset button on the Gemma itself. Enjoy your FiberBot, or skip ahead to Step
7 to find out how to create your own LED animation.

Step 6: Troubleshoot and Refine

If your hardware isn't working, first make sure no connections are broken and no wires
are touching and shorting out the circuit. Go over the steps above, to be sure everything is attached correctly.

If the software isn't running the way it should, double-check to be sure you didn't leave
out any punctuation marks. Every curly bracket and semicolon is important! Also be
sure your lines of animation code begin with a B, end with a comma, and that you have
a delay time for each frame.

Step 7: Adaptations and Extensions

Want to try your own animation? Customizing the sketch is surprisingly easy. All you
need to do is change the code on the *anim.h* page. The source code page stays the
same. Be sure to save the entire sketch under a new name.

To see how to draw your own animation frames, take a look at the animation code
itself. You'll see that—like the 8×8 LED Matrix—each frame of the animation is written
out in an 8 by 8 grid. Each LED pixel is represented by a one (indicating the light is on)
or a zero (indicating the light is off). So by looking at each block of code, you can get
an idea of what it is telling the LED Matrix to do. This is called a bitmap. (The text that
comes after the // is documentation, notes from the programmer to explain what's
happening.) Here's an example:

```
B00000000, // frame 5
B00000000,
B11111111,
B10011001,
B10011001,
B10000001,
B10000001,
B01111110,
10, // 0.10 second delay
```

As you can see, there are a few other little pieces of code you need to watch out for. Each line of code must start with a letter B and end with a comma. And after each block of eight lines representing the LED pixels is a ninth line that tells the Gemma how long that frame will last. In the example above, the number 10 followed by a comma means that frame will be displayed for a tenth (0.10) of a second. The delay time can range from 0 (no delay) to 255 (2.55 seconds). For longer delays, just repeat the same frames. According to Adafruit, the Gemma's memory can store about 320 frames of animation.

You can use any design you like for your animated LED Matrix. You can make changes right in the Arduino IDE window, or copy the EyeBlink sketch into Notebook or a word processing program, modify it to your specifications, and paste it into the *anim.h* page. The source code page stays the same. Click Save As and give your new sketch a new name. Then upload it to Gemma to see how it runs. Have fun!

Afterword: What I Learned Writing This Book

One of the things they don't emphasize enough in most books of projects for beginners is that learning to be a Maker involves failing.

Sometimes, a lot of failing.

In the course of putting together the projects in this book, I was reminded of that fact on a daily basis. Like the Beatty family with their Box of Shame, I still have the remains of projects that didn't quite work the way I wanted, or that were accidentally ruined because of a lack of skill or sheer bad luck.

Figure A-1. *It may not be pretty, but it works.*

There are also quite a few half-finished projects in my Multiple Large Storage Containers of Shame that simply grew too complicated, too expensive, or too time-consuming to include in this book. Some I may return to in the future, after I've had a chance to hone my making skills and expand my knowledge base. The rest, I'll leave to other people to solve—and use my leftover parts to build new projects.

If there's one thing I learned writing this book, it's that successful Makers—engineers, designers, artists, and programmers—have a higher tolerance for failure than the rest of us. Unexpected results are seen as a learning opportunity, and an incentive to try harder. And "Yeah, sometimes it does that, just ignore that error message" is sometimes considered an acceptable response to a question about why something isn't working.

I think that difference in attitude is one reason "beginner" projects often don't feel like they're meant for beginners. If you're used to technology working more or less reliably (and customer service taking it back and replacing it if it doesn't), then dealing with the frustration and powering through to a successful conclusion—or a better understanding of what does and doesn't work —may not come naturally.

But I believe it can be learned, like any other Maker skill. Just think back to your childhood, when *every* skill was a new one. Even basic stuff like tying knots and using scissors was a new experience, and things didn't always go as planned. But after enough practice and a little help and encouragement from those around you, eventually those skills were absorbed into the repertoire of things you didn't have to think about anymore.

With enough practice, you can master the techniques, equipment, and background knowledge needed to do the projects you want to do. And you can build up the persistence muscles that help you overcome frustration, too. If that sounds hard, I agree. I'm not there yet. But I'm trying. Every day I learn a little more, and it shows. Both in the projects I was able to create for this book, and in my interactions with other Makers.

If you're like me, you may have once felt like all you could do was walk around Maker Faire or your local makerspace and marvel at all the amazing things you saw there. But now you've taken that first step, and begun to learn some of the new techniques and technologies that have emerged since you first did arts and crafts in grade school. If you've explored the projects in this book, you can probably follow a conversation about 3D printing or Arduino that just a few months earlier would have sounded like Martian. If you've already tried them out, you're probably able to ask intelligent questions of your own.

And when you reach the point where you can answer questions from other people? That's the moment you'll feel like you've really arrived.

Throughout this book, I've suggested websites, books, and tools that can help you learn the specific background and skills you need to do the projects in this book. Here are some more general strategies:

- Start (as the Beatty family did) with kits and tutorials that lay out the materials and the directions for you.
- If at all possible, give yourself extra time, in case you have trouble figuring out a step.
- If at all possible, get extra supplies, in case something breaks or is defective.

- Scour YouTube, *makezine.com*, and other sites for videos that show you how to do things that may be hard to follow in still images.
- Ask questions of people you meet in person and in online help forums. Ignore the trolls—there are plenty of helpful folks who will take the time to get you over your roadblock.
- Learn with others at classes, workshops, and at places like makerspaces and Maker Faires in your area. Find a club or start a meet-up where you can get together with like-minded people to share ideas and collaborate.
- When you've mastered a skill, pay it forward, by helping the next person.

Good luck! And if you have questions, comments, or suggestions about this book, feel free to contact me. I'd love to hear about your adventure with making simple robots.

Kathy Ceceri

@kathyceceri (*https://twitter.com/KathyCeceri*)

www.craftsforlearning.com

For updates and news about *Making Simple Robots*, visit *craftsforlearning.com/makingsimpler-obots.htm*.

Index

Symbols

3D Modeling and Printing with Tinkercad: Create and Print Your Own 3D Models (Kelly), 72
3D printers, 50–53
 links for, 72
 on-line services for, 55
3DTin, 54
9 (film), 164

A

ABS, 51
 acetone and, 53
Absolute Beginners Guide to Building Robots (Branwyn), 103
acetone, 53
Acker, Shane, 164
Action M-Blocks, 75

actuated paper project, 4–22
 building, 9–20
 history of, 4–7
 links for, 16
 materials, 9
 mechanics of, 7
actuators, 2
Adafruit Industries, 155
 Flora microcontroler, 155
 Gemma microcontroler, 155
 help/support from, 159
 tutorials and support, 160
adhesive dots, 46
aerogel, 3
Agogino, Adrian, 37
air muscles, 24
Alice, 111
ALICE (Artificial Linguistic Internet Computer Entity), 108
Ant-Roach model, 23
Apple, 108
Ardublock, 111

Arduino microcontrollers, x
 programming, 111, 169–173
 Scratch for, 111
 sewable, 155
art by robot, 135–189
 links for, 136
 littleBits plotter project, 136–153
ArtBot, 77
artificial intelligence (AI), 105
Artificial Intelligence Markup Language (AIML), 108
artificial muscles, 2
Asimov, Isaac, vii
austenite phase of nitinol, 7
Autodesk 123D, 54, 71
automata, 73

B

Banzi, Massimo, 154, 155

batteries, 20, 158, 159

Baughman, Ray, 2

Baxter robot (ReThink Robotics), 106

Bayirli, Erkin, 37, 39

BBots (Harvard), 78

Bdeir, Ayah, 45

Beatty Robotics, x–x

Beatty, Camille, x

Beatty, Genevieve, x

Beatty, Robert, x

BeetleBot, 84

Biagini, Kevin, 50

Bicalho, Chico, 73

Big Dog robot (Boston Dynamics), 34, 127

Bio-Inspired Robotic Laboratory (National Taiwan University), 48

Blade Runner, vii

Bladerunner, 126

Block Palette (Scratch), 114

Bob the BiPed, 50

Boston Dynamics, 34

Branwyn, Gareth, 103

Breazeal, Cynthia, 105

Bristlebots, 77, 78

Brooklyn Aerodrome, 33

Buechley, Leah, 154

Buehler, William J., 7

Burton, Tim, 164

C

C-3PO, vii, 105

CAD software, 55
  Tinkercad, 56

carbon nanotubes, 2

Carnegie Mellon University, 23, 111

Case Western Reserve University, 37, 49

Cassini spacecraft, 37

Cheetah, 34

circuits, 20–21
  batteries, 158
  LED Matrix controller, 160
  links for, 21
  Ohms Law, 22
  soldering, 88–90
  wires, 158
  wiring for eFabric project, 160–169

Citilab Smalltalk Team, 111

Cleverbot chatbot (Cornell), 109

Colour Chaser, 135

Connors, Chris, 11

Cornell University, 109

Crayola Model, 129

Cubelets, 75

Cupcake CNC, 51

Curiosity rover, 37

current, 20

D

da Vinci, Leonardo, 73

DASH (Dynamic Autonomous Sprawled Hexapod), 7

Dash Robotics DIY kit, 7

DC (direct current), 20

Defense Advanced Research Projects Agency (DARPA), 35

degrees of freedom, 6

Demaine, Erik, 5, 26

Demers, Jérôme, 84

Der Kritzler (Norwegian Creations), 138

dimmer switches, 44
  in littleBits plotter, 140

DIY Mini Fume Extractor, 89

Dörfelt, Matthias, 135

Drake University, 109

Drawdio, 135

Dynalloy, 8
  how-to website, 10

voltage requirement for wires, 19

E

e-textile Arduino robot, 153–189
  Adafruit tutorials, 160
  building, 159–189
  history of, 154
  links for, 155
  materials, 156–158
  mechanics of, 155

Edgar, Emma, 33

Edgar, Marc, 33

Egg-Bot (Evil Mad Scientist), 137

Electronic Arts, Inc. (EA), 111

Eliza chatbot, 108, 109

Elmers Dry Erase Board, 144

Emma Willard Mini Maker Faire, 33

Evil Mad Scientist website, 79
  plotter, 137

EyeRoll animation, 176

F

FabLab, 138

Felt, Wyatt, 24, 25

Final Fantasy: The Spirits Within (2001), 125

Finio, Ben, 31

FIRST Robotics Competition, 33

Flexinol, 8, 9

France, Anna Kaziunas, 72

Franklin Institute, 73

Freeform Miller Solar Engine, 95–97

Fried, Limor Lady Ada, 155

frozen smoke, 1–3, 3

Fuller, Buckminster, 36

Funniest Computer Ever competition, 108

G

GeekDad, vii
GeekMom, vii
Geminoid robot (Osaka University), 125
Getting Started with Arduino (Banzi), 155
Ghazanfar, Asif, 127
Giomi, Luca, 78
GLaDOS chatbot, 122
Glue Dots, 46
Google, 34
Graffiti Research Lab (New York City), 21
graphical programming
    links for, 123
    Scratch and, 110–111
Griffith, Saul, 23

H

Hanako 2 (Yoshida Dental Manufacturing), 125
Hanks, Tom, 125
Hanson Robotics, 126
Hanson, David, 126, 126
Hart, Vi, 26
Harvard University, 5, 24
    BBots, 78
    Kilobots, 77
    RoboBee, 34
    TERMES robots, 76
Hawaii Pacific University, 108
Hayes, Gregory, 91
heat, nitinol and, 7
Hektor (Norwegian Creations), 138
Hello World programs, 122
Hexbug, 77

Hirshhorn Museum and Sculpture Garden (Smithsonian Institution), 37
hopping frog origami, 7
Hrynkiw, David, 83, 92, 103
Hugo (movie), 73
Humphrys, Mark, 109

I

I-Swarm, 78
if-then-else blocks, 119
Indiana University Department of Informatics, 126
inflatable robot project, 22–31
    building, 26
    history of, 23–24
    links for, 31
    materials, 26
    mechanics of, 24
Instructables, 51, 103
integrated development environment (IDE), 154
    installing, 170
Intel, 51
Interaction Design Institute Ivrea, 154
iPhone, 108
iRobot, 23
Ishiguro, Hiroshi, 125, 127
Isle of Man, 49

J

Jansen, Theo, 74
Jaquet-Droz, Pierre, 73
Java, 111
Jepson, Brian, 91
Jimmy (Intel), 51
Johnson, Brian David, 51
Joseph, Sam, 108
JunkBots, BugBots & Bots on Wheels (Tilden and Hrynkiw), 83, 92, 103

K

Kamen, Dean, 33
Karvinen, Kimmo, 155
Karvinen, Tero, 155
Keepon robot (Yale), 106
Kelly, James Floyd, 72
Kemp, Adam, 91
Kennedy, Erin RobotGrrl, 136
Kilobots, 77
kinetic furnishings, 4
kinetic sculptures, 4
Kismet, 105
Kogakuin University (Japan), 125

L

LabVIEW, 111
Lang, Robert J., 10
Learn How to Program with Scratch (Marji), 123
Learn to Solder (Jepson, Moskowite, Hayes), 88
LEDs, 20
    bar graph of, 44
LEGO Education WeDo kit, 111
LEGO Mindstorms, vii, 51
    programming, 111
Leonardo robot (MIT), 105
Levin, Stephen, 38
Li, David, 111
Lifelong Kindergarten group (MIT), 110, 135
LilyPad components, 154, 155
Lipson, Hod, 109
littleBits, 45
    servos, 140
littleBits plotter project, 136–153
    building, 144–152
    history of, 137
    links for, 153
    materials for, 142

mechanics of, 139
locomotion, 33–72
    degrees of freedom, 6
    links for, 35
    tensegrity robot project, 36–48
    wheel-leg hybrid robot project, 48–72
Loebner Prize, 107
Loebner, Hugh, 107
loops, 114
Loper, 49
Luxembourg, 135

M

M-Blocks, 75
MacDorman, Karl, 126
magic smoke, 91
Maillardet, Henri, 73
Make: 3D Printing: The Essential Guide to 3D Printers (France), 72
Make: Arduino Bots and Gadgets (Karvinen and Karvinen), 155
Make: Basic Arduino Projects (Wilcher), 155
Make: Electronics (Platt), 91
Make: Getting Started with Soldering Kit (Maker Shed), 88
Make: website, 10
Maker Faires, viii
Maker Movement, viii
Maker Shed website, 33
MakerBot, 50
MakerShed, 103
MaKey MaKey, 111, 136
Marji, Majed, 123
martensite phase of nitinol, 8
Matelli, Tony, 129
materials, 1–31
    3D printers and, 50–53

actuated paper project, 4–22
    inflatable robot project, 22–31
    paper robotics linkbox, 16
    shrinkable plastic, 1–3
    SMA wire, 3
McElroy, Kathryn, 154
McKibben, Joseph L., 24
MGonz, 109
Miller Solar Engine, 84, 92
Miller, Andrew, 84
Mind, 107
Minecraft, 55
MIT, 5
    RoboTuna, 34
    Step by Step Chatbot project page, 113
MIT Media Lab, 6, 10
MIT Media Labs Personal Robots Group, 105
Mitsuku chatbot, 108
Model Magic, 129
modular robots, 75
Mori, Masahiro, 124, 127
Moskowite, Tyler, 91
motors, 100–103
mountain folds, 17
MouseBot, 84
MSRox, 49

N

Nano robots, 77
NASA Ames Research Center, 37
National Taiwan University, 48
Needle Tower (Snelson), 36
New York, viii
New York City, New York, 21
Nickel Titanium Naval Ordnance Laboratory, 7
nitinol SMA wire, 3, 7
    austenite phase of, 7
    martensite phase of, 8

Nomad Press, vii
North Carolina State University, 1
Norwegian Creations, 138
Nuremburg, Germany, 138

O

Octopus Project, 3
Ohms Law, 22
origami, 4
origami robots, 5
Osaka University, 125
OtherLab, 23
    inflatable arm, 25
    website, 1
Owen, Ivan, 51

P

Pancake Bot, 51
Paul robotic arm (Tresset), 138
Pausch, Randy, 111
peel-and-stick aluminum foil tape, 9
Pettis, Bre, 39
Philip K. Dick android, 126
PLA, 51
Platt, Charles, 91
Pleo robot (Yale), 105
Plotclock (FabLab), 138, 140
plotters
    mechanics of, 140
    Traveling Salesman Problem and, 140
Pneubots, 23
Polar Express (2004), 125
Ponoko, 50, 55
pop-up illustrations, 4
potentiometer, 44
Princeton University, 127
Processing, 154